CHEMISTRY RESEARCH AND APPLICATIONS

THORIUM

CHEMICAL PROPERTIES, USES AND ENVIRONMENTAL EFFECTS

CHEMISTRY RESEARCH AND APPLICATIONS

Additional books in this series can be found on Nova's website under the Series tab.

Additional e-books in this series can be found on Nova's website under the e-book tab.

CHEMISTRY RESEARCH AND APPLICATIONS

THORIUM

CHEMICAL PROPERTIES, USES AND ENVIRONMENTAL EFFECTS

RENAE JAMISON
EDITOR

nova publishers
New York

For permission to use material from this book please contact us:
Telephone 631-231-7269; Fax 631-231-8175
Web Site: http://www.novapublishers.com

Library of Congress Cataloging-in-Publication Data

Thorium : chemical properties, uses and environmental effects / [edited by] Renae Jamison.
pages cm. -- (Chemistry research and applications)
Includes bibliographical references and index.
ISBN: 978-1-63321-309-8 (hardcover)
1. Thorium. I. Jamison, Renae, editor.
QD181.T5T46 2014
546'.422--dc23

2014021868

Published by Nova Science Publishers, Inc. † New York

CONTENTS

PREFACE

Thorium (Th) is the first element in the Actinide series having atomic number 90. It is a radioactive element. Thorium oxides can be used as nuclear fuels, although this application is quite limited compared with uranium oxide materials. In contrast to the well-known properties of bulk thorium oxides, thorium oxide molecules are much less understood. This book discusses the chemical properties, uses and environmental effects of thorium. Topics discussed include the products of laser ablated thorium atom reactions with small molecules; soil profile thorium and uranium concentration distributions in southeastern Missouri soils; influences of coal-fired power plants on the thorium levels in soils and on radioactive hazards for the population; behavior of Th-isotopes along the pigment TiO2 industrial production process; some features of the Th presence in the poços de Caldas alkaline massif, Brazil; geological and geophysical implications for western Australia; hermophysical and thermodynamic properties of oxygen-containing compounds of thorium; and unique role of thorium in closed (Th-U-Pu)-fuel cycle due to the involvement of hybrid "fusion-fission" reactors to generate advanced (231Pa+232U+233U)-fuel.

Chapter 1 – Laser ablated thorium atoms have been reacted with a variety of small molecules in our laboratory including O_2, H_2, H_2O, H_2O_2, CO, CO_2, CH_4, N_2, F_2, HF and OF_2. The authors will review the results of these earlier investigations and present new experiments and calculations on the CH_4 and HF systems where the latter features the major thorium oxide fluoride HThF reaction product. Our investigations of Th reactions with O_2 are important to verify that the ThO and ThO_2 products prepared in this reaction are the same as those evaporated from the solid at 2300°C by earlier workers. Matrix isolation investigations have produced the only examples of molecular thorium hydrides and hydroxides. In addition the first thorium methylidene complex,

CH_2=ThH_2, which exhibits agostic distortion, was prepared from the thorium reaction with methane in this laboratory. Highly favorable reactions of thorium atoms with F_2, of course, produce the stable ThF_4 molecule investigated previously, and in addition, a host of interesting new thorium fluoride intermediate species.

Chapter 2 – Understanding of the pedogenic pathways associated with Thorium (Th) and Uranium (U) migrations and accumulations in soil is (i) important to understanding Th and U soil chemistry and (ii) important for discriminating between natural background and anthropogenic concentrations. Twenty five soil series, some with multiple pedons, were assessed (i) to determine the Th and U distributions in soil profiles and (ii) to infer pedogenic processes important to the Th and U soil profile distributions. Alfisols and Ultisols were selected because they reside on stable landscapes and possess argillic horizons (diagnostic soil horizons having illuviation accumulations of clay) and possessed forest vegetation prior to European settlement. The majority of pedons exhibiting argillic horizon expression (Alfisols, Ultisols, and some Mollisols) show Fe-oxyhydroxide maxima corresponding precisely with the argillic horizon. Pearson Correlation coefficients indicate consistently statistically significant correspondences involving iron-oxyhydroxides (Fe) with both Th and U for most of the selected soils. Pedons not showing relationships involving Fe with Th and U either possess a lithologic discontinuity (change in parent material inferring inherited concentration differences) or simply lack sufficient elemental variation to formulate significant correlations. Most soils have Th/U ratios between 3 and 4; however, the range in the Th/U ratios spans from 1.4 to 12.6, demonstrating parent material heterogeneity.

Chapter 3 – Coal contains trace amounts of natural radionuclides, such as ^{232}Th. In the burning process in a coal-fired power-plant (CFPP), ashes become enriched in this radionuclide, which can be released as fly ash from the CFPP and deposed from the atmosphere on the nearby top soils, modifying the natural radioactivity background levels, and subsequently increasing the total radioactive dose received by the population. Thus, the estimation of thorium levels in the soils can be used as a tool to know the radiological impact of a CFPP. In this paper, the radiological impact of a CFPP in Spain has been studied. The activities of Thorium and other radionuclides have been measured by gamma spectrometry in coal, ash, sediment, waste and soil samples. From these results, the distribution of Thorium in soils and if deposition and migration has been taken place can be analysed taking into account its physical-chemistry properties. Analysis of Variance (ANOVA),

Principal Component Analysis (PCA), Hierarchical Clustering Analysis (HCA), and kriging mapping have been applied to the experimental dataset. Finally, the mean annual outdoor effective dose has been evaluated as 71 μSv, which is very close to the average UNSCEAR value of 70 μSv, and this means that Th concentrations found have not influence on the radioactivity levels measured in the studied area, and finally, thus confirming that the presence of the CFPP has no radioactive impact in the population.

Chapter 4 – The present study, conducted initially to characterize from radiological point of view, the raw materials (ilmenite and titanium slag), co-products (copperas and sulfafer), wastes (red gypsum and ilmenite mud) and final product (TiO_2) used/generated along the titanium dioxide industry, located in the province of Huelva, has allowed in addition a detailed study of the routes followed by several radionuclides, with special emphasis on Th-isotopes, throughout the industrial process. This pigment TiO_2 industry works by the "sulphate method", which is a typical NORM (Naturally Occurring Radioactive Materials) industry. The raw material (ilmenite, mainly formed by $FeTiO_3$), contains around 100 Bq kg^{-1} of U-series radionuclides, and around 400 Bq kg^{-1} from the Th-series ones. Along the industrial process, the secular equilibrium for the uranium and thorium series existing in the raw material is broken, and then the radionuclides are distributed/accumulated in the different materials according to their chemical behaviors. Therefore, the main goal of this study has been to gain basic information of the behavior of Th-isotopes in comparison with the behaviour of other natural radionuclides along the industrial process and, with basis in the information gained, to analyze the future potential environmental impact of co-products and wastes obtained in the industrial process. In this sense, it is important to note that some of the co-products (copperas and sulfafer) are currently used in fields such as, agriculture, fertilizers, etc.

Chapter 5 – Alkaline rocks are found in southern Brazil in association with the Ordovician-Cretaceous sedimentary Paraná basin and mainly distributed along its margins. They vary considerably in age and composition, and have been assembled into distinct geographic provinces, such as the Poços de Caldas alkaline massif, Minas Gerais State, representing one of their most important ones, particularly due to the presence of economical deposits bearing uranium, thorium and rare earth elements. Several rock types are associated with high ^{232}Th levels, mainly occurring at Morro do Ferro that is situated near the center of the Poços de Caldas plateau and initially drew the attention of mining prospectors due to the abundant occurrence of magnetite veins and dykes. It is a small hill located near the intersection of a fractures system, rising some 140 meters above the plateau base level and lying five kilometers to the north of the Osamu Utsumi uranium mine and about 10 km to the south of the city of Poços de Caldas. The superficial radioactivity in this area was systematically investigated in the middle 60s with the aim of evaluating the effects of the accentuated radioactivity in the ore body environment. The level of the environmental gamma radiation close the top of the hill varied between 1 and 3 mR/h (100-300 times higher than the background) in an area of about 30,000 m^2 and the plants were so radioactive due to the ^{228}Ra absorption that could generate *autoradiographic images.* Thorium is a highly insoluble element in water due to its presence in refractory minerals of difficult dissolution, but depending on the pH, ionic strength and others factors it may be released to the liquid phase. This chapter reports the results of some studies focusing the thorium presence in the Poços de Caldas alkaline massif with the aim of addressing some insights on its behavior in the environment.

Chapter 6 – This study reviews thorium occurrence in terms of geology and resources in Western Australia. Australia, it has been reported, has some of the highest thorium (Th) concentrations recorded in rocks in the World. High thorium concentrations have been recorded in rocks in Western Australia for over fifty years, and the association of thorium with heavy-mineral-sand deposits has been recognized during this time in the Perth and Carnarvon basins. Studies have also recognized high thorium levels in the granitoids (geological bodies with mineral composition similar to granite) of the Yilgarn and Pilbara cratons of Western Australia. Recent investigations have detected unusually high levels of thorium in granitic gneiss of the Leeuwin Complex in the southwest of Western Australia, and supported high thorium concentrations in the Northampton Complex north of the Perth Basin, which may contribute to high heat flow in these regions. The study shows known thorium distribution in Western Australia from both geological, geochemical and geophysical information. It reviews, especially based on airborne geophysics, a number of the higher thorium-concentration locations in the state and their genesis. Further, a detailed discussion of the impact of thorium on heat generation in the upper crust in Western Australia suggests that this element plays a significant role in the terrestrial heat flow in various regions, and may have implications for future geothermal energy generation. Thorium is present as a minor constituent in many minerals, but particularly monazite, and is widely scattered across Western Australia. The Department of Mines and Petroleum's database of mines, deposits and prospects in Western Australia (MINEDEX) records 110 occurrences where monazite is listed as a commodity (mineral) for a particular site. Most of these are related to heavy mineral sands along the Perth Basin, both north and south of Perth. The concentration of monazite within heavy mineral sands is highly variable, but can be up to about 3%. While there is no commercial thorium production currently in Western Australia, monazite concentrate was produced from 1963 to 1994 as a by-product from mining zircon and ilmenite. Monazite by-product is also possible from mining tantalite in pegmatite at, for example, the Wodgina and Greenbushes deposits. The presence of thorium, together with uranium and potassium, is known to contribute to heat generation within the Earth. Thorium, together with uranium (U), distributed throughout the acidic (granitoid association) rocks in the middle and upper Earth's crust, may contribute to as much as 75% of the observed heat flow out of the Earth's interior. Th^{232} has been reported to have a heat production of approximately 0.07 μWm^{-3} versus U^{238} having approximately 0.25 μWm^{-3}. Generally in Western Australia, thorium is about two to four times more abundant than uranium, which makes its contribution to heat generation almost equivalent to, and sometimes greater than, the contribution of uranium. It is estimated that, when Th/U > 3.7, then thorium will have a greater contribution to heat generation than uranium. Interestingly, recent studies in the southwest of Western Australia have shown that at some localities thorium is very much more abundant than uranium (with the ratio Th/U in the range of 10 to 15). In terms of specific concentrations in Western Australia, thorium concentrations in granitoids have been measured between 170 to over 250 ppm thorium from ground-calibrated-geophysical gamma-spectrometer surveys; corresponding uranium concentrations varied between 8 and 16 ppm. Consequently, these data appear to show that thorium can have greater importance than uranium in heat generation and heat flow in many parts of Western Australia. The comparison of surface thorium measurements with respect to recently acquired aero-radiometric and aero-magnetic data in Western Australia exhibit previously unrecognized correlations for many regions in Western Australia; some of these are reported in this review. A detailed analysis of the growing thorium database indicates a strong inter-relationship between thorium concentration, heat generation and concomitant natural terrestrial heat flow in the Leeuwin, Gascoyne and to a lesser extent in parts of the Yilgarn and Pilbara regions of Western Australia.

Chapter 7 – This chapter presents the chemical thermodynamic data set for thorium species. Analysis of the majority of calorimetric data on thorium oxide compounds is made. Thermal expansion is one of the thermophysical characteristics of the matter. It gives an information about materials behaviour in a wide range of temperature.

Chapter 8 – Technology of controlled thermonuclear fusion (CTF) is traditionally regarded as a practically inexhaustible energy source. However, development, mastering, broad deployment of fast breeder reactors and closure of nuclear fuel cycle (NFC) can also extend fuel base of nuclear power industry (NPI) up to practically unlimited scales. Under these conditions, it seems reasonable to introduce into a circle of the CTF-related studies the works directed towards solving some principal problems which can appear in a large-scale NPI in closed NFC. The first challenge is a large scale of operations in NFC back-end that should be reduced by achieving substantially higher fuel burn-up in power nuclear reactors. As was shown in [1], the use of ^{231}Pa-^{232}Th-^{232}U-^{233}U fuel in light-water reactor (LWR) opens a possibility of principle to reach very high (about 30% HM) or even ultra-high fuel burn-up. The second challenge is a potential unauthorized proliferation of fissionable materials. As is known, a certain remarkable quantity of ^{232}U being introduced into uranium fraction of nuclear fuel can produce a serious barrier against switching the fuel over to non-energy purposes. As is shown in publications [2, 3], involvement of hybrid thermonuclear reactors (HTR) into NPI structure can substantially facilitate resolving these problems. If HTR will be involved into NPI structure, then main HTR mission consists not in energy generation but in production of nuclear fuel with a certain isotope composition. The present chapter analyzes some neutron-physical features in production of advanced nuclear fuels in thorium HTR blankets. The obtained results demonstrated that such a nuclear fuel may be characterized by very stable neutron-multiplying properties during full LWR operation cycle and by enhanced proliferation resistance too. The chapter evaluates potential benefits from involvement of HTR with thorium blanket into the international closed NFC.

In: Thorium
Editor: Renae Jamison
ISBN: 978-1-63321-309-8

Chapter 1

THE PRODUCTS OF LASER ABLATED THORIUM ATOM REACTIONS WITH SMALL MOLECULES

Lester Andrews
Department of Chemistry, University of Virginia,
Charlottesville, Virginia

ABSTRACT

Laser ablated thorium atoms have been reacted with a variety of small molecules in our laboratory including O_2, H_2, H_2O, H_2O_2, CO, CO_2, CH_4, N_2, F_2, HF and OF_2. We will review the results of these earlier investigations and present new experiments and calculations on the CH_4 and HF systems where the latter features the major thorium oxide fluoride HThF reaction product. Our investigations of Th reactions with O_2 are important to verify that the ThO and ThO_2 products prepared in this reaction are the same as those evaporated from the solid at 2300°C by earlier workers. Matrix isolation investigations have produced the only examples of molecular thorium hydrides and hydroxides. In addition the first thorium methylidene complex, CH_2=ThH_2, which exhibits agostic distortion, was prepared from the thorium reaction with methane in this laboratory. Highly favorable reactions of thorium atoms with F_2, of course, produce the stable ThF_4 molecule investigated previously, and in addition, a host of interesting new thorium fluoride intermediate species.

EXPERIMENTAL AND THEORETICAL METHODS

The experimental apparatus and procedure for thorium atom reactions with small molecules have been described previously, [1-3] and a schematic diagram is illustrated in Figure 1. The Nd:YAG laser fundamental (1064 nm, 10Hz repetition rate with 10 ns pulse width) was focused onto a freshly cleaned thorium (Oak Ridge National Laboratory, high purity) target mounted on a rotating rod.

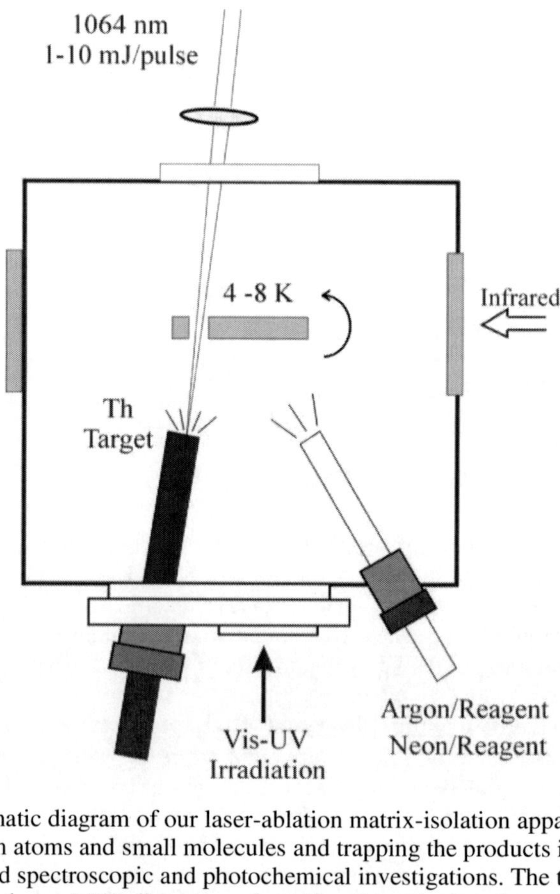

Figure 1. Schematic diagram of our laser-ablation matrix-isolation apparatus for reacting thorium atoms and small molecules and trapping the products in solid argon or neon for infrared spectroscopic and photochemical investigations. The thorium metal target is mounted on a 0.25" diameter rod rotating (1 rpm) when the pulsed 1064 laser line is focused on the target. Matrix gas containing the reagent molecule is co-deposited with Th atoms on the cold window where the reaction occurs, and the window is rotated normal to the IR beam for recording spectra. The cold window can be turned back toward the optical quartz window for sample irradiation.

Laser-ablated thorium atoms were reacted with 0.1 to 2 % of the reagent molecules in argon (Matheson, research) in all cases or neon (Spectra Gases) for selected systems using 2- 4 mmol quantities during co-deposition onto a CsI cryogenic window at 4 K or 8 K for 30 to 60 min periods. Isotopic reagents were employed when available. The cold finger was rotated 90^0 and FTIR spectra were recorded at 0.5 cm^{-1} resolution on Nicolet 550 or 750 FTIR instruments using HgCdTe range B liquid nitrogen cooled detectors. Frequency measurements were reproduced within 0.2 cm^{-1}. Matrix samples were annealed at different temperatures and cooled back to 8 K for the Nicolet 550 apparatus and to 4 K for the Nicolet 750 system for spectral acquisition. Selected samples were rotated back into the position for sample deposition and subjected to broadband photolysis through glass filters using a medium-pressure mercury arc street lamp (Philips, 175W) with the outer globe removed. After rotation of the cold window back into the infrared beam, further FTIR spectra were recorded.

Our initial DFT calculations were performed using the Gaussian 09 program system. [4] The geometries were optimized and analytical second derivatives were calculated to predict the harmonic vibrational frequencies. These calculations were performed with the B3LYP hybrid exchange-correlation functional [5] with Gaussian or aug-cc-pVTZ [6,7] basis sets on the reagent and the small core relativistic effective core potential (ECP) from the Stuttgart group with the corresponding segmented basis set on Th. [8] Additional higher level CCSD(T) and CASPT2 calculations were performed by our collaborators as described in joint publications.

Results and Discussion

We will present a brief introduction for each chemical system, followed by the results of matrix isolation infrared experiments reacting thorium atoms with selected reagent molecules, and the identification of the major reaction products. These identifications have been supported by the effect of annealing, photolysis and isotopic substitution on the infrared absorptions and by density functional and higher level electronic structure and frequency calculations in our original publications.

Thorium oxides. Thorium oxides can be used as nuclear fuels, although this application is quite limited compared with uranium oxide materials. [9-11] In contrast to the well-known properties of bulk thorium oxides, thorium oxide molecules are much less understood. Most of the experimental and theoretical studies on this subject are limited to the thorium monoxide molecule, and ThO has been characterized in both the gas phase and argon matrixes by a permanent electric dipole moment of 2.782 +/- 0.012 D in the ground X $^1\Sigma^+$ (v=0) state and vibrational frequencies at 895.8 and 876.4 cm^{-1}, respectively, in the gas phase and solid argon. [12-15] For thorium dioxide, ThO_2, absorption spectroscopic information is known only from matrix isolation studies. [14-16] Convincing infrared spectra of ThO_2 evaporated from solid thorium dioxide at 2300° C and trapped in solid argon at 15 K revealed symmetric and anti-symmetric stretching vibrational modes at 787.4 and 735.3 cm^{-1}. [14] In addition, the oxygen isotopic frequency ratio for the anti-symmetric stretching mode enabled a prediction of 122.5 ± 2° for the ∠O-Th-O angle, [14] which has stood the test of time based on more recent observations [16,17] and calculations [8c].

The element thorium is also of interest for comparing the chemistries of the group 4 transition metals and of the fellow early actinide uranium. [18,19] Group 4 metal dioxide [20-22] and tetrahydride molecules, [16,23,24] both of which have the metal in the formal +4 oxidation state, are cases in point. It is noteworthy that we find atomic thorium to be more reactive than Group 4 and U metal atoms under our experimental conditions.

Additional thorium atom reactions were done with different O_2 concentrations in argon and neon matrixes. [25] The infrared spectra from the products of laser-ablated thorium atom and 0.1% O_2 reactions revealed strong absorptions at 876.4, 787.1, and 735.1 cm^{-1} in the argon matrix. These bands have been previously assigned to the ThO molecule and the symmetric and antisymmetric O-Th-O stretching modes of ThO_2. [14-17] The ThO_2 absorptions increased four-fold when the sample was annealed to 25 K, which indicates that the Th insertion into O_2 to give the bent ThO_2 molecule is a spontaneous reaction even at very low temperatures.

$$\mathrm{Th}\ (^3F) + O_2\ (^3\Sigma_g^-) \rightarrow \mathrm{O{=}Th{=}O}\ (^1A_1)$$

In addition a new set of product bands increased at 775.7, 581.4 and 478.0 cm^{-1}, which were also present in the spectrum after sample deposition with lower intensities. No change was found for all of the new product bands when the sample was exposed to visible irradiation with a 380 nm long wavelength pass filter, but full arc ultraviolet irradiation with wavelengths longer than 220 nm substantially increased the ThO_2 absorptions. Further sample annealing markedly increased the 775.7, 581.4 and 478.0 cm^{-1} bands, which have been assigned to the thorium dioxide dimer molecule [25].

Analogous reactions were done with N_2O as the reagent in this laboratory by Kushto, and the most obvious difference was the increase in the yield of ThO relative to ThO_2, as can be reasonably expected, and the frequencies were the same within experimental error as found with the O_2 reagent. Interesting series of satellite bands on both ThO_2 bands and the N_2O absorptions are presumably due to NN complexes, which are currently under investigation.

The reactions of thorium and O_2 were also studied during condensation in excess neon to probe the influence of the matrix atoms. Table 1 summarizes the different thorium oxide frequencies observed in both argon and neon matrixes. In addition to the common oxygen species, which are always observed in these experiments due to irradiation from the laser ablation plume, [25] simple thorium oxide absorptions were observed at 887.1 (ThO), 808.7 and 756.9 cm^{-1} (ThO_2) after co-deposition with 0.5% O_2 in neon at 4 K, [16,17] and the thorium dioxide absorptions exhibited matrix site splittings, as shown in Figure 2. New product bands were also observed at 794.1, 600.7 and 487.7 cm^{-1}. The Ne-to-Ar shifts for the new product bands are less than 3%, suggesting only weak interactions between the guest molecules and matrix atoms. [26] All of the absorptions in the Ne matrix exhibited ^{18}O shifts that were similar to those observed in solid argon. The band positions are also listed in Table 1. The frequencies for the ThO_2 species in the Ne matrix of 808.7 and 756.9 cm^{-1} closely match the frequencies of 807.7 and 756.0 cm^{-1} from the high accuracy CCSD(T) calculations on the free molecule [8c,25] and show that there is a (21.7 +/- 0.1 cm^{-1}) argon matrix shift but very little shift in the neon matrix from the predicted gas phase values.

The new argon matrix absorptions at 775.7, 581.4 and 478.0 cm^{-1} exhibit identical behavior throughout the experiments, suggesting that they are different vibrational modes of the same new molecule. The 581.4 and 478.0 cm^{-1} absorptions shifted to 551.3 and 452.7 cm^{-1} for the reaction with $^{18}O_2$, the isotopic ratios of which are slightly lower than that of the diatomic ThO molecule (1.0563). A triplet splitting was observed in both mixed [$^{16}O_2$ + $^{18}O_2$] and [$^{16}O_2$ + $^{16}O^{18}O$ + $^{18}O_2$] isotopic reagent experiments, which is in line with the fact that two equivalent oxygen atoms are involved in these two modes. [25] As these band positions are too low for terminal Th=O stretching modes, we assign these transitions to Th-O stretching modes in a rhombic $(ThO)_2$ ring. The band positions for the first and last absorptions in the $^{16}O_2$ + $^{18}O_2$ precursor experiment are exactly the same as those observed in pure isotopic experiments whereas slight shifts are found for both of them in the experiment with mixed $^{16}O_2$ + $^{16}O^{18}O$ + $^{18}O_2$ sample.

Table 1. Experimental Infrared Absorptions (cm^{-1}) of Products from the Reactions of Thorium and Oxygen in Solid Ar and Ne Matrices. (Numbers in bold are Ne Matrix Frequencies)

Molecule	$^{16}O_2$	$^{18}O_2$	$^{16}O/^{18}O$ frequency ratio	$^{16}O_2$ + $^{18}O_2$	$^{16}O_2$ + $^{16}O^{18}O$ + $^{18}O_2$
ThO	876.4 [876.7][a] 887.1	829.8[830.0] 839.6	1.0562[1.0563] 1.0566	876.4, 829.8	876.4, 829.8
ThO_2	787.1, 735.1 [787.4,735.3][a] 808.7, 756.9	743.7,697.2 [..., 697.3] 763.9, 717.7	1.0584,1.0545 [.......,1.0545] 1.0583,1.0546	787.1, 735.1 743.7, 697.2	
Th_2O_4	775.7 794.1	735.9 752.8	1.0541 1.0549	776.9, 775.7	
	581.4 600.7	551.3 569.6	1.0546 1.0546	581.4, 567.0, 551.3	579.9, 566.9, 552.2
	478.0 487.7	452.7 461.7	1.0559 1.0563	478.0, 462.7, 452.7	462.8

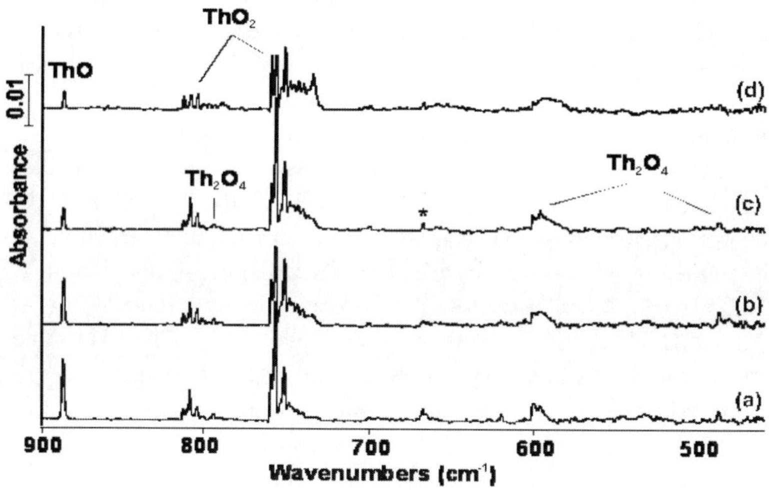

Figure 2. Infrared spectra of the laser-ablated Th atom and O_2 reaction products in solid neon at 4 K: (a) Th + 0.5 % O_2 deposition for 30 min; (b) after annealing to 8 K; (c) after $\lambda > 220$ nm irradiation; (d) after annealing to 10 K. The * denotes CO_2 inpurity.

As a result, the Th-O stretching modes in the rhombic $(ThO)_2$ ring are coupled with other vibrational modes involving oxygen atoms. Consistent with this conclusion, a new band at 775.7 cm^{-1} in the terminal Th=O stretching region was found to track with the 581.4 and 478.0 cm^{-1} absorptions, and its ^{18}O shift and isotopic $^{16}O/^{18}O$ frequency ratio, 1.0541 are slightly lower than that of the diatomic ThO molecule (1.0563). On the basis of these three experimental infrared absorptions, a symmetrical Th_2O_4 molecule containing two terminal Th=O bonds and a rhombic $(ThO)_2$ core ring is proposed, namely $O{=}Th(O)_2Th{=}O$. This structure forms readily from two bent O=Th=O molecules. Finally, it is interesting to compare the bond lengths computed at the CCSD(T) level of theory for ThO (1.845 Å) and bent OThO (1.905 Å) and measured for ThO (1.842 Å)[8c] with those calculated for the OThO dimer, $O{=}Th(O)_2Th{=}O$ (terminal 1.904 Å and ring 2.183 Å). The substantially longer ring Th-O bonds are single bonds, but the terminal bonds and those for the bent O=Th=O molecule are in the double bond length region [25].

Calculations at the CCSD(T) and B3LYP levels of theory reveal that the lowest energy structure of the Th_2O_4 molecule has a non-planar C_{2h} geometry with a closed shell singlet ground state. [25] The frequency calculations predict only three vibrational modes to be infrared active above 400 cm^{-1}. The two IR active rhombic Th-O stretches are calculated to be 611.7 and 481.1 cm^{-1} with respective $^{16}O/^{18}O$ isotopic ratios (DFT) of 1.0555 and 1.0562, which are close to the experimental frequencies of 581.4 and 478.0 cm^{-1} as well as the experimental $^{16}O/^{18}O$ ratios of 1.0546 and 1.0559. The anti-symmetric terminal Th=O stretching frequency is calculated to be 806.9 cm^{-1}, about 30 cm^{-1} higher than the 775.7 cm^{-1} band observed in the experiment. This is consistent with the moderate dipole moment determined for ThO[13] and the type of matrix shift expected from the CCSD(T) calculations on ThO_2. [25] The infrared inactive symmetric stretching mode, 805.8 cm^{-1}, is almost the same as the anti-symmetric stretching mode, as suggested from the experiments. Finally, no analogous U_2O_4 molecule has been detected in similar experiments using uranium. [25,27]. Clearly the linear O=U=O molecule would have to bend considerably to form an analogous dimer structure, and this is not a favorable reaction for the uranium dioxide molecule.

Thorium hydrides. The first spectroscopic evidence for thorium hydride molecules was our argon matrix infrared study, which reported new absorptions at 1485.2 cm^{-1} for ThH, 1480.1 and 1455.6 cm^{-1} for ThH_2, 1435.4 cm^{-1} for ThH_3, and 1444.8 cm^{-1} for ThH_4. [16] These assignments were supported by the observation of ThD_x counterparts with H/D isotopic frequency ratios between 1.4009 and 1.3999 for the vibration of a heavy metal hydride. Furthermore Th reactions were also investigated with HD and H_2 + D_2 mixtures, and the ThH and ThD bands revealed no intermediate isotopic component, as appropriate for a diatomic molecule, but two additional bands at 1467.7 and 1048.6 cm^{-1} were observed for the HThD molecule, with new bands between the symmetric and anti-symmetric stretching modes of ThH_2 and ThD_2. Additional intermediate absorptions were also found for the ThH_2D_2 molecule, which substantiated their assignments [16].

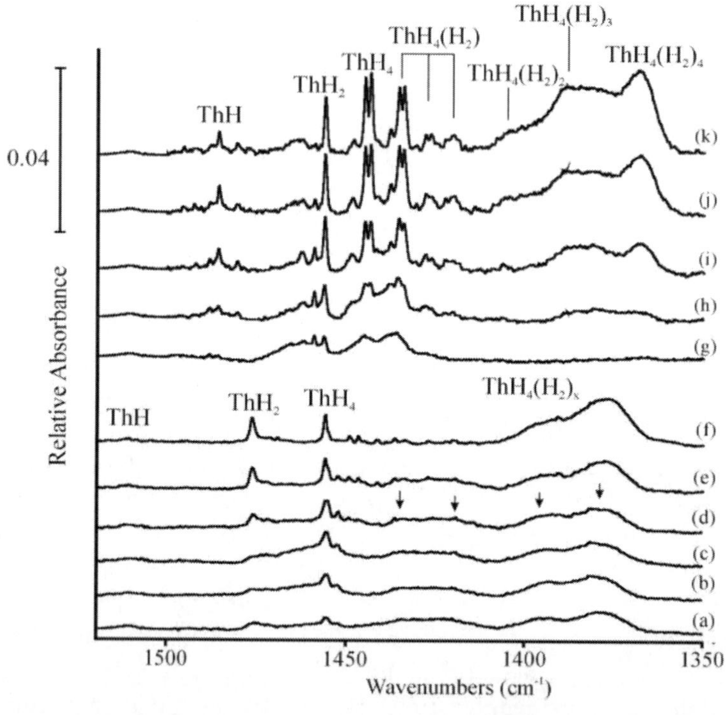

Figure 3. Infrared spectra of laser-ablated Th atoms and H_2 reaction products in solid neon and argon at 4 K: (a) Th + 3 % H_2 in neon deposition for 60 min; (b) after annealing to 7 K; (c) after 240-380 nm irradiation; (d) after annealing to 8 K; (e) after annealing to 10 K; (f) after annealing to 12 K. (g) Th + 0.5 % H_2 in argon deposition for 60 min; (h) after annealing to 30 K; (i) after annealing to 35 K; (j) after annealing to 40 K; (k) after annealing to 45 K. The arrows in (d) denote aborptions that are probably due to increasing H_2 clusters with ThH_4, namely $Th(H_2)_{1,2,3,4}$.

Our next investigation extended this work to neon and pure hydrogen matrixes, [24] and the absorptions are listed in Table 2. Argon and neon matrix spectra are compared in Figure 3.

Table 2. Observed Infrared Absorptions (cm^{-1}) of Products from the Reactions of Thorium and Hydrogen in Solid Ar, Ne and H_2 Matrices. (n. o. means not observed)

	Arg	on		Neon		Hyd	ro	gen	Ident.
H_2	HD	D_2	H_2	HD	D_2	H_2	HD	D_2	
1485.2	1485.2 1060.2	1060.2	1511.0		1078.3				ThH /ThD
1480.1	1547.7	1055.6	n. o.		n. o.				ThH_2/ ThHD /ThD_2
1455.6	1048.6	1040.3	1476.0	1497, 1062	1054.5	1478.4		1053.9	ThH_2/ ThHD /ThD_2
1444.8	1487.3, 1443.9, 1056.6, 1033.0	1032.1	1455.7	1496.2 1454.9 1063.9 1041.1	1040.1	1453.1		1029.9	ThH_4/ ThH_2 D_2/ ThD_4
			1393.1			1389.3	1412 1366	991.8	ThH_4 /$(H_2)_x$/ ThD_4 /$(D_2)_x$
			1377.7			1374.3	1008 988	979.6	ThH_4 /$(H_2)_x$/ ThD_4 /$(D_2)_x$

In neon, the absorptions blue shift to a weak band at 1511.1 cm^{-1} for ThH, only the stronger band at 1476.0 cm^{-1} for ThH_2, and to 1455.7 cm^{-1} for ThH_4. The deuterium counterparts of these features define the H/D isotopic frequency ratios 1.4014, 1.3997 and 1.3996, which are almost the same as the argon matrix ratios. We see that the 4 K experiments with hydrogen in argon trap more of the products than our earlier 7 K experiments, as expected, and the observation of a dominant band at 1374 cm^{-1} in pure solid hydrogen, with a solid deuterium component at 980 cm^{-1} and 1.4020 H/D ratio, leads us to propose that $ThH_4(H_2)_x$ clusters are being formed in these experiments. Unfortunately, our previous assignment of a 1435 cm^{-1} band to ThH_3 must be changed to the first $ThH_4(H_2)_x$ complex. [16,24] In addition, theoretical calculations show that the Th-H bond in ThH_4 is the most polarized of the group 4 metal tetrahydrides, which results in a strong dihydrogen interaction between the hydride centers and the H_2 ligands in these $ThH_4(H_2)_x$ clusters.

Thorium oxyhydrides. Subsequent investigations with thorium involved the reaction with water. [28] As expected ThO, ThO_2 and ThH_x absorptions were observed, in agreement with those discussed above. The simple new

HThO species exhibited a Th-H stretching mode at 1406.0 cm^{-1} in solid argon, slightly lower the ThH_x absorptions, and a Th-O band at 842.6 cm^{-1}, also lower than ThO at 876.4 cm^{-1}. Likewise the major product H_2ThO exhibited symmetric and anti-symmetric Th-H stretching modes at 1397.2 and 1352.4 cm^{-1}, some 83 and 103 cm^{-1} lower than ThH_2. The ThO byproduct also inserted into water to give the HThO(OH) molecule. These reactions progressed on co-deposition with laser ablated and excited Th atoms, and the yields were enhanced by subsequent ultraviolet irradiation of the matrix sample (see Figure 1).

Thorium hydroxides. Since thorium di- and tetrafluoride are stable molecules, as we will discuss later in this review, we prepared the di- and tetrahydroxide molecules through reactions with hydrogen peroxide, which can be evaporated easily from a stable urea complex. [29] Infrared spectra are illustrated in Figure 4 with a comparison of thorium reactions with H_2O_2, with a H_2O_2 and D_2O_2 mixture, and with D_2O_2 in excess argon. The new absorptions at 3750.0 and 528.1 cm^{-1} (labeled t for tetrahydroxide) and at 3737.8 and 552.6 cm^{-1} (labeled d for dihydroxide) in our lowest laser energy experiment with H_2O_2 shifted to 2766.5 and 526.8 cm^{-1} (t) or 2756.9 and 540.4 cm^{-1} (d) with D_2O_2. The sharp, weak 3750.5 cm^{-1} band in the D_2O_2 sample is from HDO_2 impurity. These bands increase substantially at the expense of H_2O_2 on annealing the sample to allow diffusion and reaction of trapped species, and the t bands increased slightly with ultraviolet irradiation while the d bands decreased slightly. Experiments with higher H_2O_2 concentrations favored the t over the d bands. Weak absorptions for HThO, DThO, H_2ThO and D_2ThO were observed as discussed above. Bands due to the HOH—O and DOD—O complexes from precursor decomposition labeled c are common to hydrogen peroxide experiments with other metals.

The experiments with $H_2O_2 + D_2O_2$ mixtures revealed broadened t bands and a new mixed isotopic satellite band at 2768.1 cm^{-1}, and we found no evidence for isotopic exchange. Finally the previous experiments with Th and H_2O were re-examined and weak d and t bands were observed (+/- 0.2 cm^{-1}) as also found with $H_2 + O_2$ reagent mixtures.

The d bands were assigned to $Th(OH)_2$ and the t bands to $Th(OH)_4$ based on the appearance of new bands in the mixed isotopic spectra for t and not d and agreement with the predictions of both MP2 and B3LYP frequency calculations that find the strong anti-symmetric O-H stretching mode for $Th(OH)_2$ to be 36 cm^{-1} below that for $Th(OH)_4$ and the strong anti-symmetric Th-O stretching mode to be 35 cm^{-1} higher for d than t. Similar relationships have been found for the group 4 metals. Our calculations converge to a slightly bent $^{1}A_1$ ground state for $Th(OH)_2$ and a tetrahedral ground state for $Th(OH)_4$.

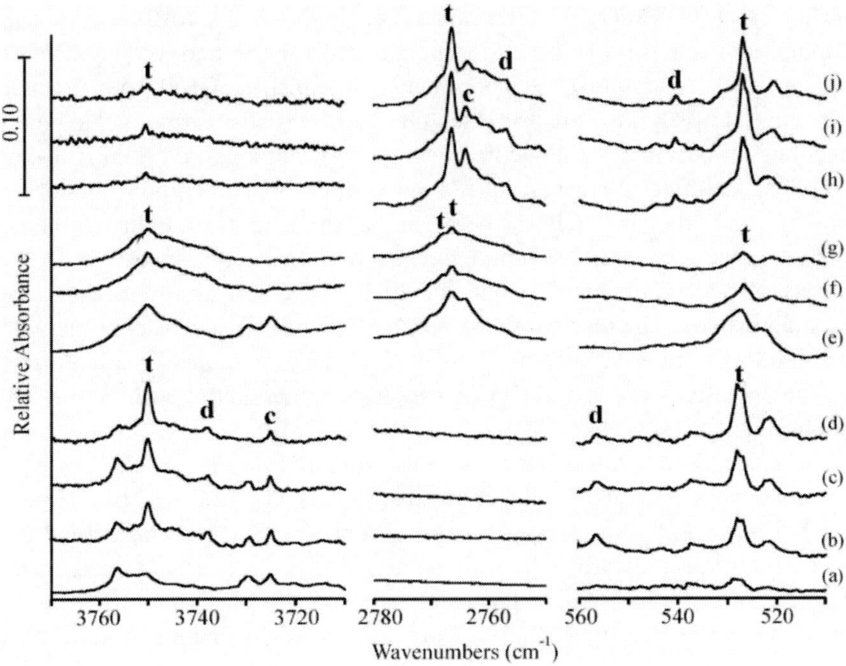

Figure 4. Infrared spectra of laser-ablated Th atom and hydrogen peroxide reaction products in solid argon at 4 K: (a) Th + H_2O_2 vapor co-deposited in excess argon for 60 min; (b) after annealing to 26 K; (c) after 240-380 nm irradiation; (d) after annealing to 34 K; (e) Th + H_2O_2 + D_2O_2 co-deposited in excess argon for 60 min; (f) after annealing to 28 K; (g) after annealing to 40 K; (h) Th + D_2O_2 co-deposited in excess argon for 60 min; (i) after annealing to 36 K; (j) after annealing to 40 K.

The reaction of a Th atom with H_2O_2 is highly exothermic, and the growth of minor d and major t absorptions on annealing shows that these reactions are spontaneous and proceed essentially without activation energy. The reaction energies computed at the MP2 level are as follows:

$$Th(^{3}F) + H_2O_2 \rightarrow Th(OH)_2\ [\Delta E = -233\ \text{kcal/mol}]$$

$$Th(OH)_2 + H_2O_2 \rightarrow Th(OH)_4\ [\Delta E = -197\ \text{kcal/mol}]$$

The monohydroxide ThOH is not observed, but the 37 kcal/mol more stable HThO isomer is produced on co-deposition of laser ablated Th atoms and H_2O_2.

ThCO and CThO. Transition metal carbonyl complexes, $M(CO)_n$, are very important subjects in modern coordination chemistry and organometallic chemistry. [19] In contrast the coordination chemistry of CO with actinide metals is a relatively new area of research, due in no small part to challenges in handling and characterizing actinide complexes. Although thorium is one of the easiest actinide elements to use in synthesis, its carbon monoxide chemistry and the differences between thorium and uranium carbonyl chemistry have not been explored extensively. We have investigated the reactions of U and CO and found an extensive chemistry with the most interesting product, the inserted CUO species, [30] which is isoelectronic with the ubiquitous uranyl dication, $[OUO]^{2+}$. Hence, we next explored the analogous reactions of Th and CO first in excess neon and then in argon and krypton [31].

The infrared spectra revealed a new band at 1817.5 cm^{-1} in the carbonyl stretching region and two weaker new absorptions at 812.2 and 617.7 cm^{-1}. Annealing the solid neon sample from 4 to 8 to 4 K more than doubled the former absorption indicating that this new carbonyl species is formed easily without activation energy, and the latter band increased perhaps 20%. Next irradiation with visible light almost destroyed the former band and more than doubled the latter two bands suggesting the photochemical rearrangement of the former to the latter product. A following full arc irradiation slightly reduced the latter bands and produced a weak former band. A final annealing to 10 K restored approximately the original former absorption but only sharpened the latter absorptions.[31]

The 1817.5 cm^{-1} band exhibits $^{13}C^{16}O$ and $^{12}C^{18}O$ counterpart frequencies at 1777.7 and 1774.3 cm^{-1}, which give 12/13 and 16/18 isotopic frequency ratios of 1.0224 and 1.0243, respectively, and are appropriate for a carbonyl vibration. These frequencies are listed in Table 3. When mixed isotopic reagents were employed, only pure isotopic counterparts were observed, which indicates a monocarbonyl complex. Therefore, these bands are assigned to the ThCO molecule. Although the mixed isotopic experiments showed that single C and O atoms are involved in the other new species, in contrast, the 812.2 and 617.7 cm^{-1} absorptions, which tracked together on photolysis, exhibited entirely different isotopic behavior. The 812.2 cm^{-1} band shifted to 811.9 cm^{-1} with $^{13}C^{16}O$ and to 769.2 cm^{-1} with $^{12}C^{18}O$, showing a small 12/13 ratio (1.0004) and a large 16/18 ratio (1.0559), and oppositely the 617.7 cm^{-1} band shifted to 595.0 with $^{13}C^{16}O$ and to 617.3 cm^{-1} with $^{12}C^{18}O$, with a large 12/13 ratio (1.0382) and a small 16/18 ratio (1.0006). This isotopic substitution clearly shows that the 812.2 cm^{-1} band is mostly Th-O and the lower 617.7 cm^{-1} band mostly Th-C in vibrational character and thusly identifies this new isomer as the CThO insertion product. Note that this ThCO >> photoisomerization << CThO is reversible with visible and near uv light [31].

Table 3. Infrared Absorptions (cm^{-1}) Observed for Products of the Reactions of Thorium and Carbon Monoxide Isotopic Molecules in Solid Ne, Ar and Kr Matrices

Molecule	$^{12}C^{16}O$	$^{13}C^{16}O$	$^{12}C^{18}O$
ThCO/Ne	1817.5	1777.7	1774.3
CThO/Ne	812.2, 617.7	811.9, 595.0	769.2, 617.3
ThCO/Ar	1787.6	1748.2	1746.2
CThO/Ar	793.1, 605.9	792.6, 583.8	751.1, 605.5
ThCO/Kr	Not obs	Not obs	
CThO/Kr	784.7, 596.0	784.7, 574.3	

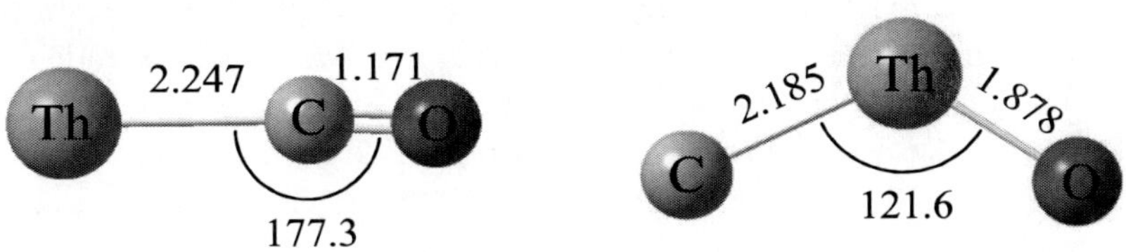

Figure 6. Structures of ThCO and CThO computed with density functional theory using the B3LYP functional with an aug-cc-pVTZ-ECP basis set.

Our Amsterdam Density Functional calculations for CThO predicted a bent (109 deg) triplet ($^3A'$) ground state 10 kJ/mol lower in energy than the triplet ($^3\Sigma^-$) ground state for ThCO and frequencies that are in excellent agreement with observed neon matrix values.[31] The calculation of a bent triplet ground state for CThO whereas CUO has a linear singlet lowest configuration is a consequence of the change in electron count from U to Th. Furthermore, the bent structure of CThO reminds us of the bent structure of ThO_2, a molecule also isoelectronic with linear CUO and uranyl, $[OUO]^{2+}$. [14,27,30,31] Cho and the author have recalculated these structures using the B3LYP hybrid density functional, the dimensions are comparable, Figure 6; however, the relative energies are reversed with ThCO being more stable by 40 kJ/mol. A more precise determination of the true lowest energy isomer will have to await higher level theoretical calculations, which are in progress in this laboratory.

It is important to examine the CThO molecule in solid argon because the CUO molecule exhibited a different, triplet, ground state in the more polarizable argon medium. [30] Isotopic spectra from Th experiments with CO, mixed $C^{16}O$, $C^{18}O$, and mixed ^{12}CO, ^{13}CO are shown in Figure 5 and the data are listed in Table 2: these unpublished spectra were recorded by B. Y. Liang in the course of other investigations in this laboratory and will be reported here. There is a band at 1787.6 cm^{-1} in the upper region, which behaves like the 1817.7 cm^{-1} neon matrix band in that it increases on annealing, but it bleaches on irradiation, and it becomes a 1787.6-1746.2 cm^{-1} doublet with mixed $^{16,18}CO$ precursor and a 1787.6-1748.2 cm^{-1} doublet with mixed $^{12,13}CO$ reagent. These isotopic frequency ratios, 1.02370 and 1.02254, are again appropriate for a C-O vibration, and as before we have a single CO absorption, and it follows that this feature is the argon matrix counterpart of the neon matrix band for ThCO described above. In the figure, we have a new 793.1 cm^{-1} band, which increases slightly on annealing and photolysis and becomes a 793.1-751.1 cm^{-1} doublet with mixed $^{16,18}CO$ precursor and shifts to 792.9 cm^{-1} with mixed $^{12,13}CO$ precursor. The 1.05592 frequency ratio is in accord with a Th-O vibration and the slight C-13 shift for the mixture, from which we can estimate the ^{13}CO counterpart as 792.7 cm^{-1}, indicates a very small mixing with the lower frequency Th-C stretching vibration. The tracking 605.9 cm^{-1} band broadens and shifts to 605.7 cm^{-1} with mixed $^{16,18}CO$ but gives a 605.9-583.8 cm^{-1} doublet with mixed $^{12,13}CO$ precursor. This ratio, 1.03786, is in line with that for a Th-C stretching mode.

An analogous experiment was done using a krypton matrix, and the lower two bands were observed at 784.7 and 596.0 cm^{-1}, but no related upper band was observed. When $^{12,13}CO$ was employed, the 784.7 cm^{-1} band was observed and the lower band gave a 596.0-574.3 cm^{-1} doublet and ratio 1.03779 as expected for a Th-C stretching mode. Hence, the 784.7 and 596.0 cm^{-1} bands are the krypton matrix counterparts for the CThO molecule. These freqencies, which are collected in Table 3, display the effect of a polarizable medium interaction and provide no evidence for a change in the ground electronic state, but they do raise a question about the relative stabilities of the two isomers in the several matrix hosts. The insertion product CThO is apparently favored in the heavier matrices as the relative yield of ThCO is less than that of CThO, and CThO increased slightly on annealing in the solid argon and krypton matrix.

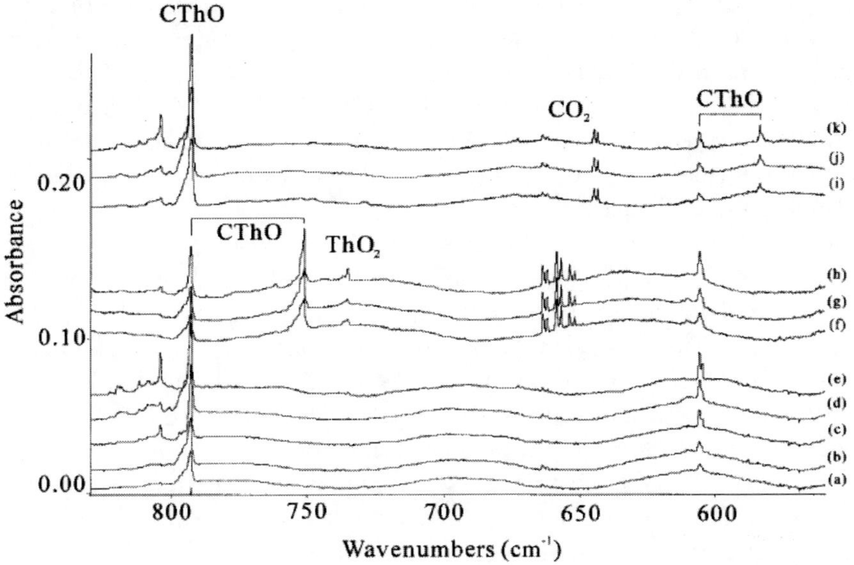

Figure 5. Infrared spectra of laser-ablated Th atom and CO molecule reaction products in solid argon: (a) Th + 0.5% CO deposition for 60 min; (b) after λ >470 nm irradiation, (c) after annealing to 25 K; (d) after λ >220 nm irradiation; (e) after annealing to 35 K ; (f) Th + 0.5% mixed $C^{16}O$, $C^{18}O$ deposition for 60 min; (g) after λ >470 nm irradiation, (h) after annealing to 25 K; (i) Th + 0.5% mixed ^{12}CO, ^{13}CO deposition for 60 min; (j) after λ >470 nm irradiation, (k) after annealing to 25 K.

Th and CO_2. The reaction of Th atoms with CO_2 forms one major new product and, of course, ThO in both the argon and neon matrix reactions. This new product absorbs at 1778.4 cm^{-1} in solid neon and at 1766.4 cm^{-1} in solid argon, a modest shift for a thorium bearing product trapped in the same electronic states in both matrixes. [26,32] The above CO isotopic precursors were employed, and the shifts were comparable, indicting again a C-O stretching mode for a single CO subunit. Recall that the carbonyl stretching mode for ThCO shifted from 1817.5 to 1787.6 cm^{-1} on going from solid neon to argon, almost two times the shift for OThCO. A simple rationale for this difference is that the Th metal center in ThCO is more exposed for interaction with the host matrix atoms than that that in OThCO.

Vibrational frequencies were calculated using the B3LYP hybrid density functional and the basis sets. The harmonic frequencies calculated for ThCO (1840.0 cm^{-1}), CThO (818.9, 532.1), and OThCO (1978.6 cm^{-1}) are higher than their previous values using the ADF density functional, [32] as generally expected.

Th and CH_4: CH_3-Th-H and CH_2=ThH_2. Since we demonstrated clearly that methylidene complexes can be prepared from group 4 metal atom reactions with methyl halides and methane, [33-35] we believed that Th would behave similarly, and following this rationale, we synthesized in solid argon the first simple, actinide methylidene complexes, CH_2=ThHF and CH_2=ThH_2, using these methods, [35,36] and the name "thora-ethylene" can be applied to the latter of these new methylidenes.

The Th-H stretching mode for CH_2=ThHF appeared at 1380.5 cm^{-1}, on the higher end of the ThH_x frequency range, [16] and the Th-F stretching mode at 518.6 cm^{-1} fell within a few cm^{-1} of ThF_4, which will be discussed below. The predominantly Th-C stretching mode was observed at 674.1 cm^{-1}, and it shifted to 674.1 cm^{-1} with C-13 and to 619.3 cm^{-1} with D substitution. The 12/13 isotopic frequency ratio, 1.03168, is lower than the value calculated for a pure harmonic diatomic Th=C stretching vibration, 1.03871, and the D shift indicates that this mode also involves the motion of H/D atoms. Hence the strongest infrared absorptions of the mehtylidene CH_2=ThHF have been identified, and the Th insertion reaction followed by alpha-hydrogen transfer is indicated [33-35].

Figure 7 shows the argon matrix infrared spectrum of the minor Th and CH_4 reaction product, noted i for insertion product hydride, and the major product labeled m for methylidene. [36,37] The stronger i band at 1359.8 cm^{-1} shows a very small 0.4 cm^{-1} shift with $^{13}CH_4$, which indicates a small coupling of the Th-H stretching vibration with a methyl group vibration. This band shifts to 970.6 cm^{-1} using the CD_4 precursor and yields an H/D ratio of 1.401, which indicates a Th-H stretching motion. Next, notice two strong m bands in the Th-H stretching region and three m bands in the lower region. The isotopic shifts for these bands characterize the vibrational modes portrayed below, and thus identify the CH_2=ThH_2 reaction product molecule. The 1435.7 and 1397.1 cm^{-1} bands clearly fall in the Th-H stretching region as our above discussion of thorium hydride vibrations indicates: these two new m bands are slightly lower that the two stretching modes for ThH_2 at 1480.1 and 1455.6 cm^{-1}. Unfortunately, their expected deuterium counterparts fall beneath the very strong CD_4 precursor absorption at 995 cm^{-1}, but they exhibit no C-13 shifts. The next band at 670.8 cm^{-1} shifts to 651.5 cm^{-1} with $^{13}CH_4$, which gives the 12/13 frequency ratio 1.0296, a number that is short of the 1.0387 value predicted for a harmonic diatomic Th=C vibrational mode. Note that this vibration falls near the Th=C mode, 674.1 cm^{-1}, for CH_2=ThHF discussed above. [35] A comparison with the lower 605.9 cm^{-1} Th=C stretching mode for OThC, which exhibits a 1.0379 12/13 isotopic frequency ratio almost exactly the diatomic Th=C stretching 1.03871 value. [31] The deuterium shift to 602.9 cm^{-1} shows that this mode is also coupled to a motion involving H/D, but the ratio 1.1126 is way short of that for any C-H bending mode, so the 670.8 cm^{-1} band can be called a mostly Th=C stretch interacting with a C-H bending mode. The 634.6 cm^{-1} band shows the small ^{13}C shift (5.4 cm^{-1}) and larger D shift (ratio 1.27) appropriate for a CH_2 wagging mode [36].

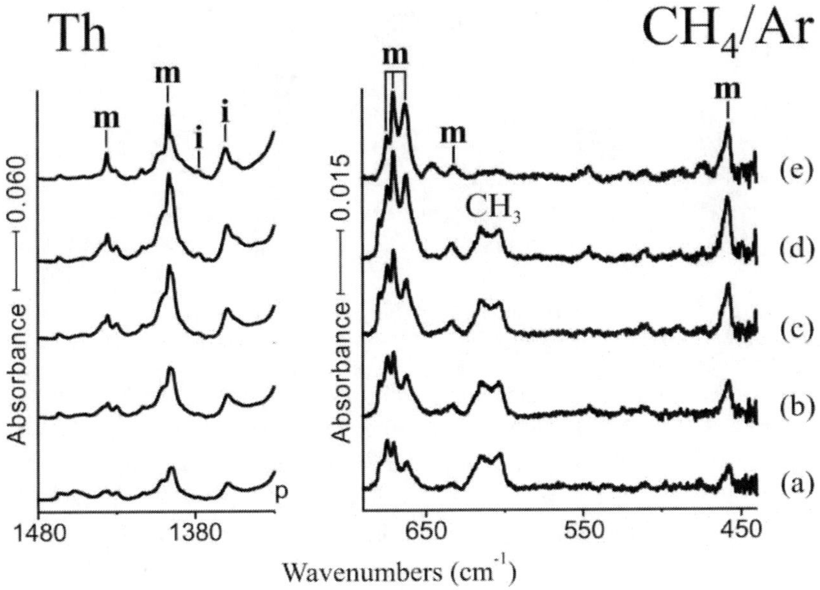

Figure 7. Infrared spectra of laser-ablated Th atom and CH_4 molecule reaction products in solid argon: (a) Th + 2% CH_4 in argon deposited for 60 min onto 8 K CsI window; (b) after λ >420 nm irradiation, (c) after 240-380 nm irradiation; (d) after λ >220 nm irradiation; (e) after annealing to 26 K in sequence for the same matrix sample. CH_4 precursor absorption is noted p.

New B3LYP density functional calculations were performed by H.-G. Cho for CH_2=ThH_2 using the basis sets employed for our recent Th_2O_4 calculations, [8c, 25] and the CH_2=ThH_2 structure is illustrated in Figure 8 below. These dimensions compare favorably with those from our earlier calculations using the B3LYP density functional and the CCSD method with Gaussian basis sets [36].

Conversion of methane to methanol is important since methanol is both a fuel as well as a starting material for numerous synthetic reactions. [38] We investigated the raction of Th and CH_4 in the presence of O through the ablation of thoria ceramic, ThO, and found that the methane to methanol coversion may be catalyzed by thorium oxide through the $CH_3Th(O)H$ intermediate in solid argon [39].

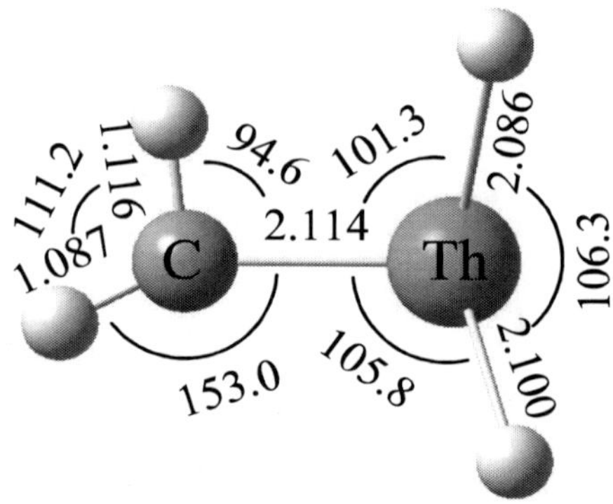

Figure 8. Structure of the methylidene CH_2=ThH_2 computed with density functional theory using the B3LYP functional with an aug-cc-pVTZ-ECP basis set [see ref. 25].

Thorium nitrides. Reactions of Th and N_2 have been performed under argon dilution and in neat nitrogen. [40] In the first case isolated ThN and N=Th=N have been observed at 934.3 and 756.6 cm^{-1} in solid argon, and the $Th(NN)_2$ and $Th(NN)_4$ complexes have been observed at 1895 and 2080 cm^{-1}. Spectra for these molecules are compared in Figure 8 for mixtures of $^{14}N_2$ and $^{15}N_2$ [trace (a)] and statistical 25% $^{14}N_2$ and 50 % $^{14}N^{15}N$ and 25 % $^{15}N_2$ [trace(b)] all diluted in argon to 0.5%. Our B3LYP calculations find a linear NNThNN molecule and a tetrahedral $Th(NN)_4$ higher species. Using the simple mixture of $^{14}N_2$ and $^{15}N_2$ the linear molecule gives the anti-symmetric stretching modes of the pure isotopic molecules at 1895 and 1831 cm^{-1} plus both symmetric and anti-symmetric modes at 1948 and 1852 cm^{-1} for the lower symmetry 14-14-Th- 15-15 isotope. The statistical mixture produces a dectet absorption for the strong anti-symmetric stretching mode and three weak bands with observable intensity for mixed isotopic derivatives in the symmetric stretching mode. In neat solid nitrogen the spectrum is dominated by a very strong 2099 cm^{-1} band for $Th(NN)_4$ $[NN]_x$ including an unknown number of dinitrogen ligands.

Thorium fluorides. Reactions of laser ablated Th atoms co-deposited with molecular fluorine (F_2) are illustrated in Figure 10 for 0.5 % F_2 (scans a-c) and 1.0% F_2 (scans d-g). [41] The strong absorptions at 521.0, 519.2 and 515.5 cm^{-1} in the region of the 520 cm^{-1} gas phase absorption are in agreement with argon matrix spectra of ThF_4 vapor from the solid at 800-850 °C. [42] New absorptions of interest here are the unresolved doublet at (575.9, 575.1) cm^{-1}, the resolved doublet at (567.2, 564.8) cm^{-1}, the 531.0 cm^{-1} band, and the 567.5 and 528.4 cm^{-1} bands which survive 20 K annealing, decrease on > 220 nm irradiation, and increase slightly on final 30 K annealing. Meanwhile a band at 460 cm^{-1} appears on deposition, decreases on 20 K annealing, increases slightly and sharpens on > 220 nm irradiation, and decreases again on final 30 K annealing. The F_3^- absorption at 510 cm^{-1} is destroyed by > 220 nm irradiation. [45] Note the reversal of the 531.0 and 528.4 cm^{-1} band intensities with increasing F_2 concentration, where more F_2 favors the lower frequency band.

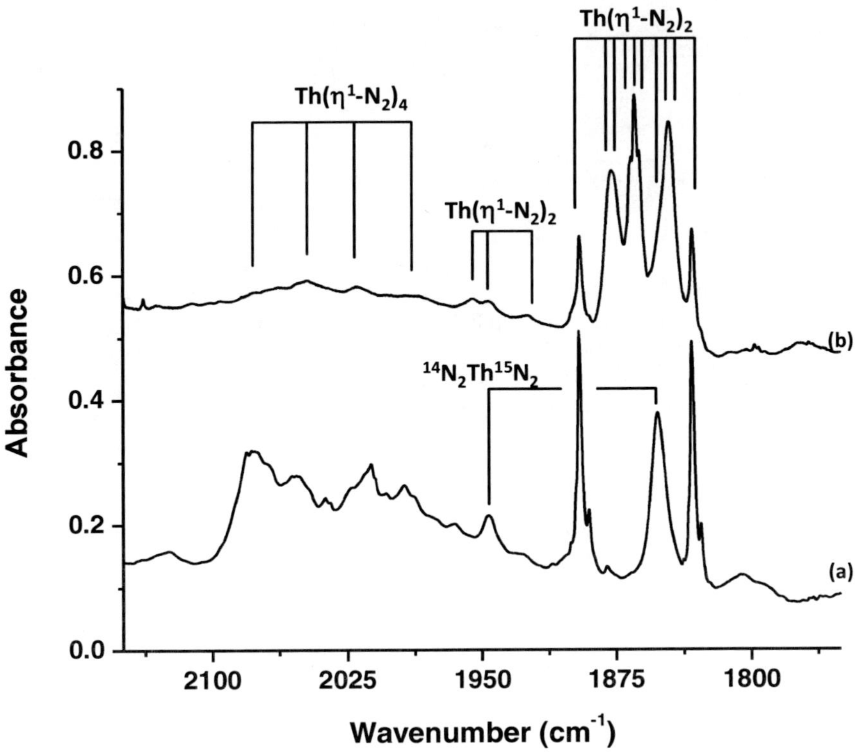

Figure 9. Infrared spectra of laser-ablated Th atom and mixed isotopic N_2 reaction products in solid argon at 4 K: (a) Th + 50 % $^{14}N_2$ and 50 % $^{15}N_2$ at 0.5% in argon deposition for 30 min; (b) Th + 25 % $^{14}N_2$ and 50 % $^{14}N^{15}N$ and 25 % $^{15}N_2$ at 0.5% in argon deposition for 30 min.

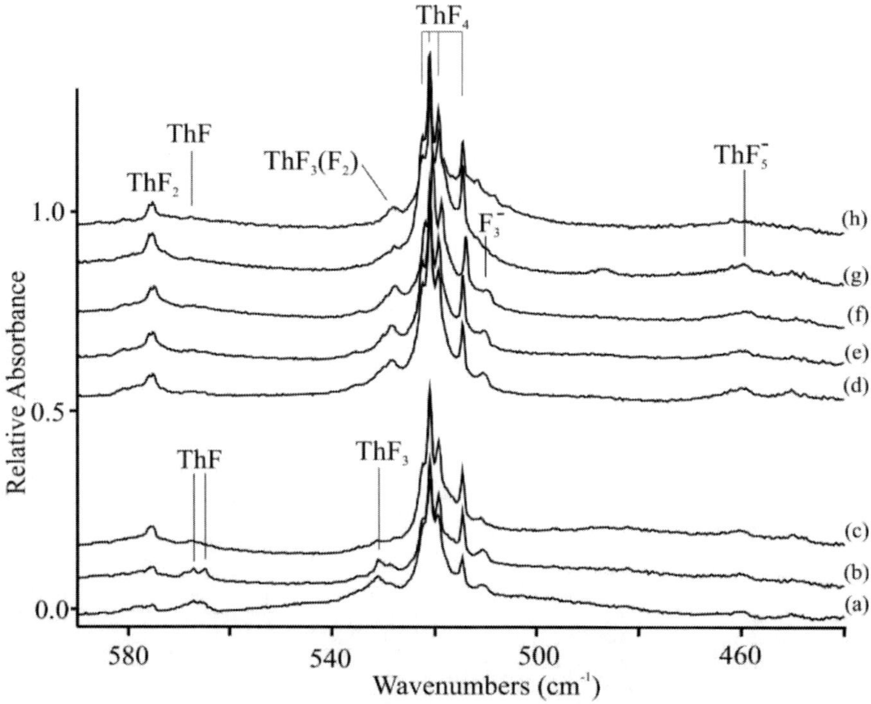

Figure 10. Infrared spectra of laser-ablated Th atom and F_2 molecule reaction products in solid argon: (a) Th + 0.5% F_2 deposition for 60 min; (b) after annealing to 20 K; (c) after λ >220 nm irradiation [spectra absorbance scale expanded x 2]; (d) Th + 1% F_2 deposition for 60 min; (e) after annealing to 20 K; (f) after λ >220 nm irradiation; (g) after annealing to 30 K.

The new absorptions at (567.2, 564.8) cm^{-1}, (575.9, 575.1) cm^{-1} and (531.0 and 528.4) cm^{-1} in argon are assigned to ThF, ThF_2, ThF_3, and $ThF_3(F_2)$ on the basis of their behavior on sample concentration, annealing, and irradiation, the use of NF_3 and OF_2 precursors, and comparison with frequencies calculated at the DFT/B3LYP and CCSD(T) levels with an aug-cc-pVTZ-ECP basis set, which are discussed in detail in our longer paper. [44] The additional broad band at 460 cm^{-1} has been assigned to the ThF_5^- anion. The trigonal bipyramidal ThF_5^- anion (calculated electron detachment energy of 7.17 eV) increases at the expense of $ThF_3(F_2)$ and the isolated F_3^- anion [45] on full mercury arc irradiation. The ionic complex $[ThF_3^+][F_2^-]$ is shown by calculations to contain a side-bound F_2^- subunit. This work reported the first evidence for thorium pentafluoride species including the unique $[ThF_3^+][F_2^-]$ ionic radical ion pair molecule and its electron capture product, the very stable ThF_5^- anion, and provided the first example of pentavalent thorium [41,44].

Thorium hydride fluoride. Reactions of laser ablated Th atoms with molecular HF gave small yields of the above $ThF_{1,2,3,4}$ molecules along with HF_2^- at 1377.0 cm^{-1} plus two new absorptions at 570.2 and 1464.7 cm^{-1}. [41,45,46] The latter bands increased on annealing to 25 K and were destroyed by >220 nm photolysis. DF enriched experiments provided new bands shifted to 568.0, 1046.0 and DF_2^- at 965.8 cm^{-1}. The isolated bifluoride anion absorptions have been produced by argon resonance vacuum ultraviolet irradiation of argon matrix samples of HF and DF. [45] The 1464.7/1046.0 = 1.4003 isotopic frequency ratio is near that for thorium hydrides discussed above, and the 1464.7 cm^{-1} band is in the middle of the Th-H stretching region. The small deuterium shift in the 570.2 cm^{-1} band to 568.0 cm^{-1} indicates a minor interaction with an H/D motion, and the 570.2 cm^{-1} band is a few cm^{-1} above the ThF diatomic molecule vibrational mode mentioned in the last section. Hence, the new absorptions at 570.2 and 1464.7 cm^{-1} are assigned to the two stretching modes of the HThF molecule, which is calculated to have a bent ground state of $^1A'$ symmetry [46].

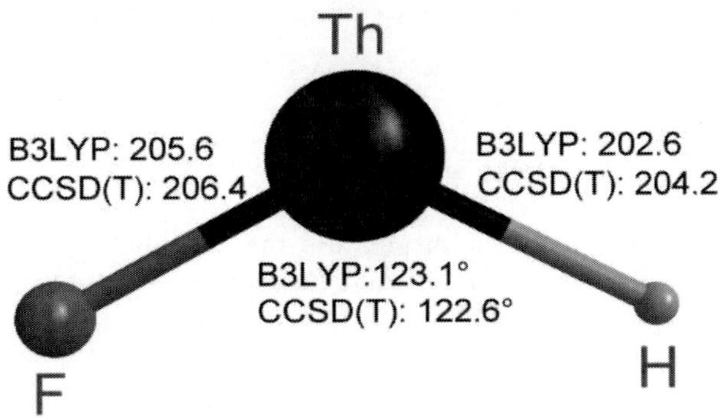

Figure 11. Structure computed for thorium hydride fluoride using two methods: bond lengths are given in pm [ref. 46].

Thorium oxide difluoride. Laser ablated Th atoms react with OF_2 to form the terminal oxo $OThF_2$ major product species via the spontaneous and specific insertion and rearrangement reaction, and also $ThF_{1,2,3,4}$ minor products. [47] The isolated $OThF_2$ molecule is characterized by one terminal Th-O and two F-Th-F stretching vibrational modes observed in the matrix isolation infrared spectrum at 834.8, 522.2, and 487.2 cm^{-1}, which are further supported by density functional frequency calculations and CASPT2 energy and structure calculations. [47] The major product has a pyramidal structure and singlet ground state. The molecular orbitals and metal-oxygen bond lengths for $OThF_2$ indicate triple bond character for the terminal oxo bond. Dative bonding involving $O_{2p} \rightarrow Th_{6d}$ interaction is clearly involved in this oxoactinide difluoride molecule. Finally, the weak O-F bond in OF_2 as well as the strong Th-O and Th-F bonds make the reaction to form the $OThF_2$ product specific and highly exothermic.

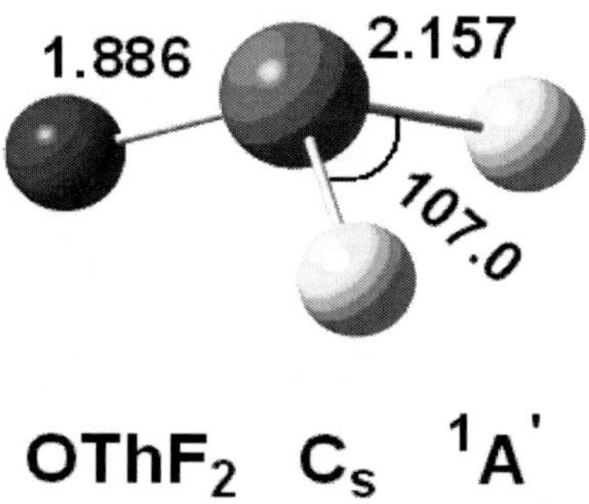

Figure 12. Structure calculated for thorium oxide difluoride using the B3LYP/6-311+G(d) level of theory [ref. 47].

ACKNOWLEDGMENTS

The author acknowledges with pleasure the contributions of co-workers whose names are given in the references. Several former collaborators have helped with the production of figures, some from their unpublished work in this laboratory, and this assistance from B. Liang, H.-G. Cho, and G. P Kushto is extra praiseworthy. We gratefully acknowledge earlier financial support from the NSF and the DOE, when many of these investigations were performed, and "retirement" support from TIAA/CREF for L.A. since 2008.

REFERENCES

[1] Andrews, L.; Citra, A. Infrared Spectra and Density Functional Theory Calculations on Transition Metal Nitrosyls. Vibrational Frequencies of Unsaturated Transition Metal Nitrosyls. *Chem. Rev.* 2002, *102*, 885-911.

[2] Andrews, L. Matrix Infrared Spectra and Density Functional Calculations of Transition Metal Hydrides and Dihydrogen Complexes. *Chem. Soc. Rev.* 2004, *33*, 123-132.

[3] Andrews, L.; Cho, H.-G. Matrix Preparation and Spectroscopic and Theoretical Inventigations of Simple Methylidene and Methylidyne Complexes of Group 4-6 Transition Metals, *Organometallics,* 2006, *25*, 4040-4053. (Invited review).

[4] Frisch, M. J.; Trucks, G. W.; Schlegel, H. B.; Scuseria, G. E.; Robb, M. A.; Cheeseman, J. R.; Scalmani, G.; Barone, V.; Mennucci, B.; Petersson, G. A.; Nakatsuji, H.; Caricato, M.; Li, X.; Hratchian, H. P.; Izmaylov, A. F.; Bloino, J.; Zheng, G.; Sonnenberg, J. L.; Hada, M.; Ehara, M.; Toyota, K.; Fukuda, R.; Hasegawa, J.; Ishida, M.; Nakajima, T.; Honda, Y.; Kitao, O.; Nakai, H.; Vreven, T.; Montgomery, Jr., J. A.; Peralta, J. E.; Ogliaro, F.; Bearpark, M.; Heyd, J. J.; Brothers, E.; Kudin, K. N.; Staroverov, V. N.; Kobayashi, R.; Normand, J.; Raghavachari, K.; Rendell, A.; Burant, J. C.; Iyengar, S. S.; Tomasi, J.; Cossi, M.; Rega, N.; Millam, N. J.; Klene, M.; Knox, J. E.; Cross, J. B.; Bakken, V.; Adamo, C.; Jaramillo, J.; Gomperts, R.; Stratmann, R. E.; Yazyev, O.; Austin, A. J.; Cammi, R.; Pomelli, C.; Ochterski, J. W.; Martin, R. L.; Morokuma, K.; Zakrzewski, V. G.; Voth, G. A.; Salvador, P.; Dannenberg, J. J.; Dapprich, S.; Daniels, A. D.; Farkas, Ö.; Foresman, J. B.; Ortiz, J. V.; Cioslowski, J.; Fox, D. J. *Gaussian 09*, Revision B.01; Gaussian, Inc.: Wallingford, CT, 2010.

[5] Becke, A. D. *J. Chem. Phys.* 1993, *98*, 5648-5652. (b) Lee, C.; Yang, W.; Parr, R. G. *Phys. Rev. B* 1988, *37*, 785-789.

[6] Frisch, M. J.; Pople, J. A.; Binkley, J. S. *J. Chem. Phys.* 1984, *80*, 3265-3269.

[7] Kendall, R. A.; Dunning, T. H. Jr. Harrison, R. J. *J. Chem. Phys.* 1994, *96*, 6796-6806.

[8] Cao, X.; Dolg, M.; Stoll, H. Valence Basis Sets for Relativistic Energy-Consistent Small-Core Actinide Pseudopotentials. *J. Chem. Phys.* 2003, *118*, 487-496; (b) Cao, X.; Dolg, M. *J. Molec. Struct. (Theochem)*, 2004, *673*, 203-209. (c) Jackson, V. E.; Craciun, R.; Dixon, D. A.; Peterson, K. A.; de Jong, W. A. Prediction of Vibrational Frequencies of UO_2^{2+} at the CCSD(T) Level (also ThO_2 and basis sets for Th). *J. Phys. Chem. A* 2008, *112,* 4095-4099.

[9] Lung, M.; Gremm, O. *Nucl. Eng. Des.* 1998, *180*, 133-146.

[10] Bae, K.-M.; Kim, M.-H. *Nucl. Eng. Tech.* 2005, *37*, 91-97.

[11] Lombardi, C.; Luzzi, L.; Padovani, E.; Vettraino, F. *Prog. Nucl. Energy* 2008, *50*, 944-953.

[12] Heaven, M. C. Probing Actinide Electronic Structure using Fluorescence and Multi-Photon Ionization Spectroscopy. *Phys. Chem. Chem. Phys.* 2006, *8*, 4497-4509.

[13] Wang, F.; Le, A.; Steimle, T. C.; Heaven, M. C. The permanent electric dipole moment of thorium monoxide, ThO. *J. Chem. Phys.* 2011, *134*, 031102.

[14] Gabelnick, S. D.; Reedy, G. T.; Chasanov, M. G. Infrared Spectra and Structure of Some Matrix-Isolated Lanthanide and Actinide Oxides. *J. Chem. Phys.* 1974, *60*, 1167-1171.

[15] Gong, Y.; Zhou, M. F.; Andrews, L. *Chem. Rev.* 2009, *109*, 6765-6808.

[16] Souter, P. F.; Kushto, G. P.; Andrews, L.; Neurock, M. Experimental and Theoretical Evidence for the Isolation of Thorium Hydride Molecules in Argon Matrices. *J. Phys. Chem. A* 1997, *101*, 1287-1291. (including ThO and ThO_2 absorptions).

[17] Zhou, M. F.; Andrews, L. Infrared Spectra and Pseudopotential Calculations for NUO^+, NUO, and NThO in Solid Neon. *J. Chem. Phys.* 1999, *111*, 11044-11049.

[18] Pyykko, P. *Chem. Rev.* Relativistic Effects in Structural Chemistry. 1988, *88*, 563-594.

[19] Cotton, F. A.; Wilkinson, G.; Murillo, C. A.; Bochmann, M. *Advanced Inorganic Chemsitry*, 6th edn., Wiley: New York, 1999.

[20] Chertihin, G. V.; Andrews, L. Reactions of Laser Ablated Titanium, Zirconium and Hafnium Atoms with Oxygen Molecules in Condensing Argon. *J. Phys. Chem.* 1995, *99*, 6356-6366.

[21] Zheng, W.; Bowen, K. H., Jr.; Li, J.; Dąbkowska, I.; Gutowski, M. *J. Phys. Chem. A* 2005, *109,* 11521-11525.

[22] Li, S.; Dixon, D. A., *J. Phys. Chem. A*, 2010, *114*, 2665-2683.

[23] Chertihin, G. V.; Andrews, L. Reactions of Laser Ablated Ti Atoms with Hydrogen during Condensation in Excess Argon. Infrared Spectra of the TiH, TiH_2, TiH_3, and TiH_4 Molecules. *J. Am. Chem. Soc.* 1994, *116*, 8322-8327. (b) Chertihin, G. V.; Andrews, L. Infrared Spectra of ZrH_4 and HfH_4 in Solid Argon. *J. Am. Chem. Soc.* 1995, *117*, 6402-6403. (c) Chertihin, G. V.; Andrews, L. Reactions of Laser-Ablated Zr and Hf Atoms with Hydrogen. Matrix Infrared Spectra of the MH, MH_2, and MH_4 Molecules. *J. Phys. Chem.* 1995, *99*, 15004-15010.

[24] X. Wang, L. Andrews, and L. Gagliardi, "Infrared Spectra of ThH_2, ThH_4, and the Hydride Bridging $ThH_4(H_2)_x$ Complexes in Solid Argon, Neon, and Hydrogen," *J. Phys. Chem. A* 2008, *112,* 1754-1761.

[25] Andrews, L.; Gong, Y.; Liang, B.; Jackson, V. E.; Flamerich, R.; Li, S.; Dixon, D. A. Matrix Infrared Spectra and Predicted Properties of Thorium Oxide Species. *J. Phys. Chem. A* 2011, *115*, 14407-14416.

[26] Jacox, M. E. The Vibrational Energy Levels of Small Transient Molecules Isolated in Neon and Argon Matrices *Chem. Phys.* 1994, *189*, 149-170.

[27] Zhou, M. F.; Andrews, L.; Ismail, N.; Marsden, C. Infrared Spectra of UO_2, UO_2^+, and UO_2^- in Solid Neon. *J. Phys. Chem. A* 2000, *104*, 5495-5502.

[28] Liang, B. Y.; Andrews, L.; Li, J.; Bursten, B. E. Experimental and Theoretical Studies of the Products of Laser-Ablated Thorium Atom Reactions with H_2O in Excess Argon. *J. Am. Chem. Soc.* 2002, *124*, 6723-6733.

[29] Wang, X. F.; Andrews, L. Infrared Spectra and Structures of the $Th(OH)_2$ and $Th(OH)_4$ Molecules. *Phys. Chem. Chem. Phys.* 2005, *7*, 3834 - 3838.

[30] Zhou, M. F.; Andrews, L.; Li, J.; Bursten, B. E. Reactions of Laser-Ablated Uranium Atoms with CO: Infrared Spectra of the CUO, CUO^-, OUCCO, $(\eta^2$-$C_2)$ UO_2, and $U(CO)_x$ (x=1-6) Molecules in Solid Neon. *J. Am. Chem. Soc.* 1999, 121, 9712-9721. (b) Li, J.; Bursten, B. E.; Liang, Andrews, L. Noble Gas-Actinide Compounds: Complexation of the CUO Molecule by Ar, Kr, and Xe Atoms in Noble Gas Matrices. *Science* 2002, *295*, 2242-2245. (c) L. Andrews, L.; Liang, B.; Li, J.; Bursten, B. E. Noble Gas-Actinide Complexes of the CUO Molecule with Multiple Ar, Kr, and Xe Atoms in Noble-Gas Matrices. *J. Am. Chem. Soc.* 2003, *125*, 3126-3139.

[31] Zhou, M. F. Zhou; Andrews, L.; Li, J.; Bursten, B. E. Reactions of Th Atoms with CO: The First Thorium Carbonyl Complex and an Unprecedented Bent Triplet Insertion Product. *J. Am. Chem. Soc.* 1999, *121*, 12188-12189.

[32] Andrews, L.; Zhou, M. F.; Liang, B.; Li, J.; Bursten, B. E. Reactions of Laser-Ablated U and Th with CO_2: Neon Matrix Infrared Spectra and Density Functional Calculations of OUCO, OThCO, and Other Products. *J. Am. Chem. Soc.* 2000, *122*, 11440-11449.

[33] Cho, H.-G.; Wang, X. F.; Andrews, L. The C-H Activation of Methane by Laser-Ablated Zirconium Atoms: CH_2=ZrH_2, the Simplest Carbene

Hydride Complex, Agostic Bonding, and $(CH_3)_2ZrH_2$. *J. Am. Chem. Soc.* 2005, *127*, 465-473.

[34] Cho, H.-G.; Andrews, L. Preparation of CH_3TiF and $(CH_3)_2TiF_2$ from the Reaction of CH_3F with Laser-Ablated Ti Atoms*Inorg. Chem.* 2004, *43*, 5253-5257. (CH_2=TiHF).

[35] Lyon, J. T.; Andrews, L. Formation and Characterization of Thorium Methylidene CH_2=ThHX Complexes. *Inorg. Chem.* 2005, *44*, 8610. (CH_2=ThHF).

[36] Andrews, L. Cho, H.-G. Infrared Spectrum and Structure of CH_2=ThH_2. J. *Phys. Chem. A* 2005, *109*, 6796-6798.

[37] Cho, H.-G.; Andrews, L. J. *Phys. Chem. A* 2014, *118*, to be submitted. (CH_3-ThH, and CH_2=ThH_2).

[38] Holmen, A. *Catal. Today* 2009, *142*, 2-8.

[39] Gong, Y.; Andrews, L.; Jackson, V. E.; Dixon, D. A. Methane to Methanol Conversion Catalyzed by ThO through the $CH_3Th(O)H$ Intermediate. *Inorg. Chem.* 2012, *51*, 11055-11060.

[40] Kushto, G. P.; Souter, P. F.; Andrews, L. An Infrared Spectroscopic and Quasirelativistic Theoretical Study of the Coordination and Activation of Dinitrogen by Thorium and Uranium Atoms. *J. Chem. Phys.* 1998, *108*, 7121-7129.

[41] Andrews, L.; Thanthiriwatte, K. S.; Wang, X. F.; Dixon, D. A. Thorium Fluorides ThF, ThF_2, ThF_3, ThF_4, $ThF_3(F_2)$, and ThF_6^- Characterized by Infrared Spectra in Solid Argon and Electronic Structure and Vibrational Frequency Calculations. *Inorg. Chem.* 2013, 52. 8228-8233.

[42] Buchmarina, V. N.; Gerasimov, A. Yu.; Predtechenskii, Yu. B.; Shklyarik, V. G. Vibrational Spectra of CeF_4 and ThF_4 Molecules Isolated in Low-Temperature Matrices. *Optics and Spectrosc. (USSR)* 1992, *72,* 38-40.

[43] Riedel, S.; Köchner, T.; Wang, X. F.; Andrews, L. Polyfluoride Anions, a Matrix-Isolation and Quantum-Chemical Investigation. *Inorg. Chem.* 2010, *49*, 7156-7164.

[44] Thanthiriwatte, K. S.; Wang, X. F.; Andrews, L.; Dixon, D. A.; Metzger, J.; Vent-Schmidt, T.; Riedel, S. Properties of ThF_x from Infrared Spectra in Solid Argon and Neon with Supporting Electronic Structure and Thermochemical Calculations. J. *Phys. Chem. A* 2014, *118*, 2107–2119.

[45] McDonald, S. A. Andrews, L. Infrared Spectrum and Structure of the Isolated HF_2^- Ion in Solid Argon. *J. Chem. Phys.* 1979, *70*, 3134-3136.

[46] Vent-Schmidt, T.; Metzgar, J.; Andrews, L.; Riedel, S. To be submitted for pub. (HThF).

[47] Gong,Y.; Wang, X. F.; Andrews, L.; Schlöder, T.; Riedel, S. Infrared Spectroscopic and Theoretical Investigations of the OUF_2 and $OThF_2$ Molecules with Triple Oxo Bond Character. *Inorg.Chem.* 2012, *51*, 6983–6991.

In: Thorium
Editor: Renae Jamison
ISBN: 978-1-63321-309-8

Chapter 2

SOIL PROFILE THORIUM AND URANIUM CONCENTRATION DISTRIBUTIONS IN SOUTHEASTERN MISSOURI SOILS

***Michael Aide**[*,1]**, Donn Beighley**[1] **and David Dunn**[2]*

[1]Department of Agriculture, Southeast Missouri State University, Cape Girardeau, Missouri

[2]University of Missouri-Fisher Delta Research Center, Portageville, Missouri

ABSTRACT

Understanding of the pedogenic pathways associated with Thorium (Th) and Uranium (U) migrations and accumulations in soil is (i) important to understanding Th and U soil chemistry and (ii) important for discriminating between natural background and anthropogenic concentrations. Twenty five soil series, some with multiple pedons, were assessed (i) to determine the Th and U distributions in soil profiles and (ii) to infer pedogenic processes important to the Th and U soil profile distributions. Alfisols and Ultisols were selected because they reside on stable landscapes and possess argillic horizons (diagnostic soil horizons having illuviation accumulations of clay) and possessed forest vegetation prior to European settlement. The majority of pedons exhibiting argillic

* Corresponding author, M.T. Aide, mtaide@semo.edu, Southeast Missouri State University, 1 University Plaza, Cape Girardeau, MO 63701

horizon expression (Alfisols, Ultisols, and some Mollisols) show Fe-oxyhydroxide maxima corresponding precisely with the argillic horizon. Pearson Correlation coefficients indicate consistently statistically significant correspondences involving iron-oxyhydroxides (Fe) with both Th and U for most of the selected soils. Pedons not showing relationships involving Fe with Th and U either possess a lithologic discontinuity (change in parent material inferring inherited concentration differences) or simply lack sufficient elemental variation to formulate significant correlations. Most soils have Th/U ratios between 3 and 4; however, the range in the Th/U ratios spans from 1.4 to 12.6, demonstrating parent material heterogeneity.

Keywords: Aqua-regia digestion, argillic horizon, cambic horizon, thorium, uranium

INTRODUCTION

Nuclear Chemistry of Thorium and Uranium

In 1828, J.J. Berzelius isolated thorium (Th) from the Norwegian ore "thorite". In 1841 B. Peligot, using a reductive process, produced metallic uranium. Chadwick discovered the neutron in 1932, which eventually lead to the understanding that neutron capture and subsequent β^- emission yielded an element with one additional proton. In 1940 E.M. McMillian and P.H. Abelson demonstrated that neutron capture converted $^{238}U_{92}$ to $^{239}U_{92}$; after which, β-emission yielded $^{239}Np_{93}$. G.T. Seaborg and E.M. McMillan (1951) shared the Nobel Prize for their discovery of numerous elements of the Actinide Series.

Thorium (Th) is the first element in the Actinide series having atomic number 90 and an electronic configuration of [Rn] $6d^2 7s^2$. Uranium is the third element in the Actinide series having an atomic number of 92 and an electronic configuration of [Rn] $5f^3 6d^1 7s^2$. The 5f orbitals are less effective in penetrating the inner core electrons than the 4f orbitals (Lanthanide series) and are, therefore, more energetically favored for bonding and may exhibit appreciable covalent character. The second element in the Actinide series is Protactinium, having two isotopes and an exceptionally trace crustal abundance. Thorium(IV) has an ionic radius of 95 pm, whereas uranium(IV) and uranium(VI) have ionic radii of 89 and 73 pm, respectively. The long lived isotope of thorium is $^{232}Th_{90}$ and the two long-lived isotopes of uranium are $^{235}U_{92}$ and $^{238}U_{92}$.

Thorium ($^{238}Th_{92}$) is a radioactive element, decaying to $^{228}Ra_{88}$ by α-emission, followed by β^- emission to yield $^{228}Ac_{89}$ and a second β^- emission to yield another isotope of thorium ($^{228}Th_{90}$). $^{228}Th_{90}$ decays by α-emission to yield $^{224}Ra_{88}$. The half-life of $^{232}Th_{90}$ decay to ^{228}Ra is long-lived (1.41 x 10^{10} years), whereas the half-lives of the other elements in the sequence are comparatively short (hours to several years). Uranium decay is an isotope function, with (i) $^{238}U92$ decaying by α-emission to $^{234}Th_{90}$ (half life of 4.45 x 10^9 years) and by two successive β- emissions to yield $^{230}Th_{90}$ and (ii) $^{235}U_{92}$ decaying by α-emission to yield $^{231}Th_{90}$ (half life of 7.04 x 10^8 years) and later in the decay sequence to yield $^{227}Th_{90}$.

Introduction to Soil Thorium and Uranium

The earth's crustal abundance of thorium is approximately 8.1 mg Th/kg, whereas uranium's crustal abundance is approximately 2.3 mg U/kg (Greenwood and Earnshaw. 1984). Rock types vary greatly in their thorium concentrations, with granites (16 mg Th/kg) and shales (12 mg Th/kg) having greater abundances than basalts (2.7 mg Th/kg) (Wanty and Nordstrom, 1995). Similarly, rock types vary greatly in their uranium concentrations, with granites (4.4 mg U/kg) and shales (3.8 mg U/kg) having greater abundances than basalts (0.8 mg U/kg) (Wanty and Nordstrom, 1995). Thus, neither element is particularly rare. In the case of thorium, geologic processes tend not to accumulate thorium in specific formations and thorium remains rather disperse. Thorium is largely associated with the mineral 'monazite', (lanthanum, other rare earth elements, Th-bearing phosphate). The high density and relative inertness of monazite to chemical weathering permits monazite accumulation in fluvial sediments or as monazite sands. Thorium is also associated with the mineral 'uranothorite' [$(U,Th)SiO_4$].

Common uranium bearing minerals include: uraninite [UO_2], pitchblende [U_3O_8], coffinite [$U(SiO_4)_{1-x}(OH)_{4x}$], brannerite [UTi_2O_6], davidite [$(REE)(Y,U)(Ti,Fe^{3+})20O_{38}$], and thucholite [Uranium-bearing pyrobitumen]. Uranium-bearing minerals that are less common include: autunite [$Ca(UO_2)_2(PO4)_2$ x 8-12 H_2O], carnotite [$K_2(UO_2)_2(VO_4)_2$ x 1–3 H_2O], seleeite [$Mg(UO_2)_2(PO_4)_2$ x 10 H_2O], torbernite [$Cu(UO_2)_2(PO_4)_2$ x 12 H_2O], tyuyamunite [$Ca(UO_2)_2(VO_4)_2$ x 5-8 H_2O], uranocircite [$Ba(UO_2)_2(PO_4)_2$ x 8-10 H_2O], uranophane [$Ca(UO_2)_2(HSiO_4)_2$ x 5 H_2O], and zeunerite [$Cu(UO_2)_2(AsO_4)_2$ x 8-10 H_2O]. Uranium(V) species and associated minerals are comparatively rare because of disproportionation into U(IV) and U(VI) species (Greenwood and Earnshaw, 1984).

Soil Chemistry of Thorium

Thorium hydrolysis, with emphasis on the normal soil pH environment ranges of pH 3 to pH 8, has been reviewed (Baes and Mesmer, 1976; Hichung, 1989; Langmuir, 1997). The predominant monomer Th-OH species include: Th^{4+}, $Th(OH)^{3+}$, $Th(OH)_2^{2+}$, $Th(OH)_3^{+}$, and $Th(OH)_4$. Polymeric Th-OH species include: $Th_2(OH)_2^{6+}$, $Th_2(OH)_3^{5+}$, and $Th_2(OH)_4^{4+}$, $Th_4(OH)_8^{8+}$, $Th_6(OH)_{15}^{9+}$, and $Th_6(OH)_{14}^{10+}$. The more important thorium hydrolysis reactions are:

$$Th^{4+} + H_2O = Th(OH)^{3+} + H^+ \qquad \log K_{11} = -3.2$$

$$Th^{4+} + 2H_2O = Th(OH)_2^{2+} + 2H^+ \qquad \log K_{12} = -6.93$$

$$Th^{4+} + 3H_2O = Th(OH)_3^{3+} + 3H^+ \qquad \log K_{13} = -11.7$$

$$Th^{4+} + 4H_2O = Th(OH)_4^{3+} + 4H^+ \qquad \log K_{14} = -15.9$$

$$2Th^{4+} + 2H_2O = Th_2(OH)_2^{6+} + 2H^+ \qquad \log K_{14} = -6.14$$

The log K hydrolysis values for $Th_4(OH)_8^{8+}$ and $Th_6(OH)_{15}^{9+}$ are log $K_{4,8}$ = -21.1 and log $K_{6,15}$ = -36.76, respectively. The speciation of the important thorium species as a function of pH is illustrated in Figure 1.

The solubility of ThO_2 may be estimated from thermochemical data as:

$$ThO_2 + 4H^+ = Th^{4+} + 2H_2O \qquad \log_{Ks1,0} = 6.6$$

Thorium readily forms complexes with fluoride (ThF_2^{2+} and ThF_3^{+} in strongly acidic media), phosphate ($ThH_2PO_4^{3+}$ near pH 1 and $Th(HPO_4)_2$ with optimum speciation at pH 4.6 and $Th(HPO_4)_3^{2-}$ with optimum speciation at pH 6), and sulfate ($ThSO_4^{2+}$ and $Th(SO_4)_2$ at pH 1) (Langmuir and Herman, 1980). Thorium also forms strong coordinate bonds with oxalate, citric acid and ethylenediaminetetraacidic acid (Langmuir and Herman, 1980). Sulfate, fluoride, phosphate and the organic ligands were shown to increase the solubility of Thorianite (ThO_2) (Langmuir and Herman, 1980). ThO_2 and noncrystalline $Th(OH)_4$ have been proposed as the solid phases most likely to limit Th solubility, with an expected range of 10^{-10} to 4×10^{-4} mole/liter in low-Eh ground waters and 10^{-10} to 10^{-7} mole/liter in high-Eh ground waters (Langmuir, 1997).

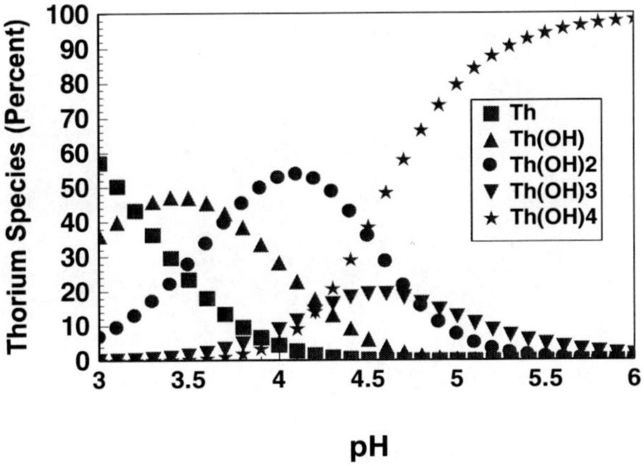

Figure 1. The distribution of important Th species as a function of pH.

The adsorption of Th onto substrates has garnished recent attention. Riese (1982) assumed the adsorption of Th was according to:

$$SOH + Th^{4+} + nH_2O = SOTH(OH)n + (n+1)H^+,$$

where SOH was the site-bonding hydroxyl and n has values of 1, 2, 3, and 4. Riese observed that adsorption on quartz (α-SiO2) with Na_2SO_4 as a supporting electrolyte (ionic strength = 0.01 mole/liter) was optimized from pH 3 to pH 4. Sulfate complexes limited thorium adsorption.

Soil Chemistry of Uranium

Hydrolysis of U(IV) is presented as (Baes and Mesmer, 1976):

$$U^{4+} + H_2O = U(OH)^{3+} + H^+ \qquad \log K_{1,1} = -0.65$$

$$U^{4+} + 2H_2O = U(OH)_2^{2+} + 2H^+ \qquad \log K_{1,2} = -0.2.6$$

$$U^{4+} + 3H_2O = U(OH)_3^{+} + 3H^{+} \qquad \log K_{1,3} = -5.8$$

$$U^{4+} + 4H_2O = U(OH)_4 + 4H^{+} \qquad \log K_{1,4} = -10.3$$

$$U^{4+} + 5H_2O = U(OH)_5^{-} + 5H^{+} \qquad \log K_{1,5} = -16.0$$

$$6U^{4+} + 15H_2O = U_6(OH)_{15}^{9+} + 15H^{+} \qquad \log K_{6,15} = -17.2$$

The solubility of U(IV) may be estimated from thermochemical data, with the assumption that UO_2 is the crystalline phase, as:

$$UO_2 + 4H^{+} = U^{4+} + 2H_2O \qquad \log K_{s1,0} = -1.8$$

or

$$UO_2 + 2H_2O + OH^{-} = U(OH)_5^{-} \qquad \log K_{s1,5} = -3.77.$$

Hydrolysis of U(VI) is presented as (Baes and Mesmer, 1976):

$$UO_2^{2+} + H_2O = UO_2(OH)^{+} + H^{+} \qquad \log K_{1,1} = -5.8$$

$$2UO_2^{2+} + 2H_2O = (UO_2)_2(OH)_2^{+} + 2H^{+} \qquad \log K_{2,2} = -5.62$$

$$3UO_2^{2+} + 5H_2O = (UO_2)_3(OH)_5^{+} + 5H^{+} \qquad \log K_{3,5} = -15.63$$

In low ionic strength media, the U(VI) polymers are not as thermodynamically favored, with the exception of = $(UO_2)_3(OH)_5^{+}$.

The solubility of U(VI) may be estimated from thermochemical data, with the assumption that $UO_2(OH)_2$ is the crystalline phase (Baes and Mesmer, 1976) as:

$$UO_2(OH)_2 + 2H^{+} = UO_2^{2+} + 2H_2O \qquad \text{Log } K_{s1,0} = 5.6.$$

The reduction of U(VI) to U(IV) may be presented as (Baes and Mesmer, 1976):

$$UO_2^{2+} + 4H^{+} + 2e^{-} = U^{4+} + 2H_2O \qquad E^{o} = 0.329 \text{ volts.}$$

Using MinteqA2 software (Allison et al., 1991), U(VI) speciation may be estimated from thermochemical data for pH intervals from pH 4 to pH 8. Setting the total U(VI) concentration at 10^{-8} mole/liter, the pCO_2 pressure at 0 and again at 0.02 bar were the primary model inputs. Establishing a constant ionic strength with 0.01 mole $NaNO_3$/liter, activity coefficients were estimated using the Davis equation. In the CO_2 closed system, UO_2^{2+} is the dominant species in very acidic media, whereas $UO_2(OH)^+$ predominates from pH 6 to pH 8 (Table 1). The ion pair $UO_2NO_3^+$ is an important secondary species, particularly in acidic media. In the CO_2 open system, UO_2^{2+} is the dominant species in very acidic media; however, the UO_2CO_3, $UO_2(CO_3)_2^{2-}$, and $UO_2(CO_3)_3^{4-}$ are increasingly dominant U(VI) species upon transition from acidic media to neutral and then alkaline media. Importantly, the uranyl carbonate complexes are stable at Eh conditions that would promote U(VI) reduction in CO_2 closed systems. This MinteqA2 simulation of dilute U(VI) speciation corresponds closely with the analytical data and its MinteqA2 simulation as presented by Langmuir (1997) and also the data analysis from Waite et al. (1994).

Table 1. Estimated uranium(VI) speciation (percent) using MinteqA2 with and without carbon dioxide spanning a pH interval from pH 4 to pH 8

CO_2 at 0.0 bar					
U(VI) species	4	5	6	7	8
UO_2^{2+}	97.8	90.4	50.8	5.5	trace
$UO_2(OH)^+$	trace	8.4	47.2	51.3	14.0
$(UO_2)_2(OH)_2^{2+}$	trace	trace	trace	trace	1.1
$UO_2NO_3^+$	2.2	1.2	trace	trace	trace
$(UO_2)_3(OH)_5^+$	trace	trace	trace	42.0	85.7
CO_2 at 0.02 bar					
U(VI) species	4	5	6	7	8
UO_2^{2+}	96.7	43.9	trace	trace	trace
$UO_2(OH)^+$	trace	4.1	trace	trace	trace
$UO_2NO_3^+$	1.3	trace	trace	trace	trace
UO_2CO_3	1.1	51.3	84.2	4.5	trace
$UO_2(CO_3)_2^{2-}$	trace	trace	14.1	78.9	2.6
$UO_2(CO_3)_3^{4-}$	trace	trace	trace	16.6	97.4

Typically, the pH range of minimal solubility of the U(VI) minerals is also the pH range for optimal adsorption of U(VI). In general, U(IV) complexes are less soluble and less mobile than U(VI) complexes (Phillips et al., 2006). Hsi and Langmuir (1985) investigated the adsorption of U(VI) onto noncrystalline $Fe(OH)_3$ and goethite (α-FeOOH) in 0.1 mole $NaNO_3$/liter suspensions as a function of total carbonate and pH. His and Langmuir documented that the optimum adsorption pH was near pH 6.3 to 6.5 for noncrystalline $Fe(OH)_3$ and U(VI)-carbonate complexes reduce U(VI) adsorption in alkaline media. For goethite suspensions the pH of maximum U(VI) adsorption shifted from pH 5.7 to pH 8.0 as the total carbonate concentration decreased. Waite et al. (1994) investigated U(VI) adsorption onto hydrous ferric oxides and documented that the maximum U(VI) adsorption spanned the range from pH 5 to pH 9, with increased carbonate concentrations limiting the U(VI) adsorption in the pH interval from pH 8 to pH 9. In general, U(VI) adsorption into Fe-oxyhydroxides is greater than phyllosilicate minerals. At Eh values less than 0.2 volts, U(VI) reduction to uraninite (UO_2) is favored. Stewart et al. (2007) observed that U(VI) reduction is inhibited in the presence of ferrihydrite.

Precipitation reactions involving U(VI) include (Chen and Yiacoumi, 2002):

$$UO_2^{2+} + 2H_2O = UO_2(OH)_2 + 2H^+ \qquad \log K = -5.4$$

$$UO_2^{2+} + CO_3^{2-} = UO_2CO_3 \qquad \log K = 14.11$$

$$2UO_2^{2+} + Ca^{2+} + 2PO_4^{3-} = Ca(UO_2)_2(PO_4)_2 \qquad \log K = 48.61$$

$$2UO_2^{2+} + Fe^{2+} + 2PO_4^{3-} = Fe(UO_2)_2(PO_4)_2 \qquad \log K = 46.00$$

$$UO_2^{2+} + H^+ + PO_4^{3-} = H(UO_2)PO_4 \qquad \log K = 25.00$$

Great interest exists to better understand Th and U transport in natural systems, e.g. soil profiles, sediments and aquifers (Aide et al., 2011; Bidoglio et al., 1989; Hamby et al., 2002; Varinlioglu and Kose, 2005; Johnson et al., 2004). Johnson et al. (2004) investigated Nevada soil sites having been impacted by depleted uranium. Uranium retention is based primarily on soil type, soil binding site concentrations, the presence of phyllosilicates and their associated Fe-oxyhydroxides, the contaminant concentration, the presence of competing ions, and the pH and Eh dependent contaminant speciation. The observed that distribution coefficients, Kd=concentration of the sorbed contaminant/the contaminant in the aqueous phase) increased with soil pH 7 to pH 11. The presence of U-silicate mineral aggregates limited vertical vadose zone transport. Roh et al. (2000) investigated two U impacted sites at Oak Ridge, TN. Sequential leaching demonstrated that soil U was substantially associated with carbonates (45%) and Fe-oxyhydroxides (40%).

Thorium and Uranium Pathways in the Soil Environment

Plant uptake of Th and U has been investigated in pristine and impacted regions to (i) infer the extent of plant uptake and (ii) determine the likelihood that plant materials may be employed to remove Th and U from the soil environment (Ebbs et al., 1998; Raju and Raju, 2000; Shahander and Hossner, 2002; Morton et al., 2002; Shtangeeva and Ayrault, 2004; Kumar et al., 2008). Harvest removal is the withdrawal of Th- and U-bearing plant material from the soil environment, whereas residues (litter) are plant materials placed at the soil surface by mechanical harvesting or natural forest/grassland accumulation (Figure 2). In India, Kumar et al. (2008) compared Th and U uptake in plant materials with respect to total soil abundances, observing that U, relative to Th, was preferentially accumulated by plant materials. Morton et al. (2002) observed the uptake of Th and U in Blueberries (*Uaccinium pallidum* Aiton) grown in Spodosols. They documented mean blueberry Th concentrations in the leaf (1.59×10^{-1} mg Th/kg) and stem (9.1×10^{-2} mg Th/kg) was substantially greater than the corresponding U concentrations (8.65×10^{-3} mg U/kg for leaf tissues and 7.95×10^{-3} mg U/kg for stem tissues). Shtangeeva and Ayrault (2004) investigated Th uptake in wheat (*Triticum vulgare* var 'Horst') in soil not considered Th or U contaminated and observed that wheat seedlings accumulated 0.1 mg Th/kg in the leaves and 0.71 mg Th/kg root systems. They speculated that microbial activity and exuded organic acids in the rhizosphere may have promoted Th plant uptake.

Residue decomposition generally results in the return of inorganic Th and U to the soil environment. Soil organic matter (humus) is a highly reactive semi-stable product of the microbially-directed soil decomposition of plant and animal tissues. Soil organic matter contains variable concentrations of aromatic and aliphatic carboxyl, alcohol and amine groups whose resultant charge density is pH dependent (Norvell, 1991). At a pH of 7, soil organic matter has a negative charge density, which retains cations, including Th and U (Chopping and Shambhag, 1981). Organic complexes of Th and U may be replaced by other cations, especially divalent and trivalent cations, buffering the soil solution (Essington, 2004).

Phyllosilicates (clay minerals) typically manifest a net negative charge density because of isomorphic substitution and unsatisfied edge charges. These negative charges provide charge capacity to electrostatically retain cations, including U and Th (Moon et al., 2006). Additionally, Mn- and Fe-oxyhydroxides are variable charges surfaces that acquire a positive charge density when the pH is more acid than the mineral's point of zero net charge density (Sato et al., 1997; Chen and Yiacoumi, 2002;). Thus Th and U, along with its hydroxyl-monomers and hydroxyl-polymers, participate in adsorption reactions at the colloidal interfaces of the phyllosilicates and Mn- and Fe-oxyhydroxides (Barnett et al., 2000 and 2002; Cygan, 2002; Davis et al., 2002; Honeyman and Ranville, 2002; McKinley et al., 1995; Runde, 2002; Runde et al., 2002; Turner et al., 1996). Wind and water erosion of clays and other colloids may be an important source of U and Th transport (Bednar et al., 2004).

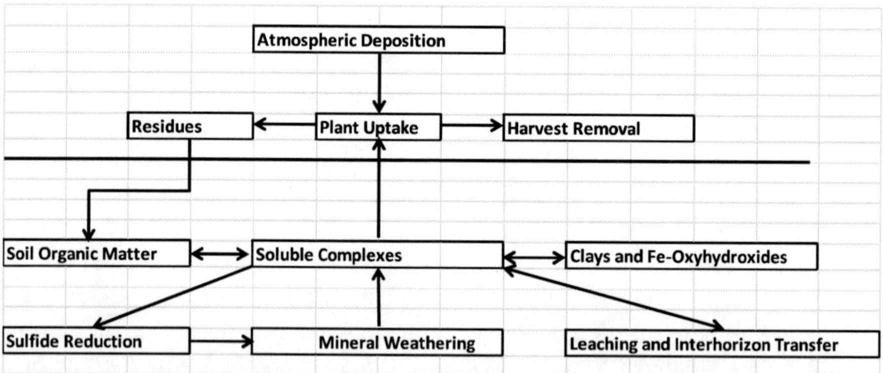

Figure 2. Illustration of the various U and Th pathways in the soil environment.

The objective of this research was to determine the soil profile distributions of thorium and uranium to infer the soil processes responsible for their accumulation in diagnostic soil horizons. All soils are considered pristine, that is, they have not been impacted by U and Th because of human activities.

MATERIALS AND METHODS

Soil profiles were selected to represent many soil series having an appreciable areal extent across southeastern Missouri and which also provided a broad array of soil series having alluvial and loess parent materials (Table 1 and Table 2). The studied soil orders include: Alfisols, Ultisols, Mollisols, Entisols, Inceptisols and Vertisols. Several of the soils (Caneyville, Frenchmill, Hildebrecht, Irondale, Kilarney, and Knobtop series) are located near ancestral Pb-mine sites, whose Pb-ore was known to have secondary concentrations of Zn and Cd (Aide et al., 2008, Aide 2009, Aide and Braden, 2009).

All soils were classified and sampled in excavated pits using USDA-NRCS protocols (Soil Survey Division Staff, 1993). Routine laboratory analysis include mechanical analysis, cation exchange capacity, pH, exchangeable cations, total acidity, Bray-1 phosphorus, 2M KCl sulfate extraction, and organic matter by loss on ignition using routine methods (Carter, 1985).

The soil characterization data was used primarily to verify that the pedons actually represented the soil series and to assess if these soil properties show relevant relationships with the Th and U soil profile distributions.

The Selected Soils of the Study Area

Table 2 displays the classification and drainage of soils in the study area, whereas Table 3 displays the parent materials and the existing vegetation/usage. The pedons range from excessive well-drained to very poorly-drained. The Alfisols, Ultisols and Mollisols have A (ochric, umbric or mollic) - E – Bt (argillic or natric) – C or R horizon sequences. The Entisols and Inceptisols possess Ap (ochric) – Bw (cambic) – C horizon sequences. The Vertisols possess Ap – Bssg horizon sequences. Soil textures range from loamy sand to clay.

Summers are hot and humid with a mean July temperature of 26°C and winter temperatures are mild with a mean January temperature of 2°C. The mean annual precipitation of 1.19 m is seasonally distributed, with greater rainfall in spring. Vegetation is either a mixed hardwood forest or land that has been cleared of vegetation, land-graded and employed in row-crop agriculture. The study area is remote from any thorium or U industrial or mining areas and is considered a relatively pristine soil resource in terms of Th and U.

Table 2. Classification and drainage class of soils

Soil Series	Classification	Drainage Class
Alred	Loamy-skeletal over clayey, siliceous, semiactive, mesic Typic Paleudalfs	well drained
Amagon	Fine-silty, mixed, active, thermic Typic Endoaqualfs	poorly drained
Broseley	Loamy, mixed, superactive, thermic Arenic Hapludalfs	somewhat excessively drained
Calhoun	Fine-silty, mixed, active, thermic Typic Glossaqualfs	poorly drained
Caneyville	Fine, mixed, active, mesic Typic Hapludalfs	well drained
Clana	Mixed, thermic Aquic Udipsamments	moderately well drained
Commerce	Fine-silty, mixed, superactive, nonacid, thermic Fluvaquentic Endoaquepts	somewhat poorly drained
Crowley	Fine, smectitic, thermic Typic Albaqualfs	somewhat poorly drained
Dubbs	Fine-silty, mixed, active, thermic Typic Hapludalfs	well drained
Foley	Fine-silty, mixed, active, thermic Albic Glossic Natraqualfs	poorly drained
Frenchmill	Loamy-skeletal, mixed, active, mesic Typic Paleudults	well drained
Heldebrecht	Fine-silty, mixed, active, mesic, Oxyaquic Fragiudalfs	moderately well drained
Irondale	Loamy-skeletal, mixed, active, mesic Typic Hapludults	well drained
Killarney	Loamy-skeletal, mixed, active, mesic Typic Fragiudults	moderately well drained
Knobtop	Fine-silty, mixed, active, mesic Aquic Hapludults	moderately well drained
Lilbourn	Coarse-loamy, mixed, superactive, nonacid, thermic Aeric Fluvaquents	somewhat poorly drained
Malden	Mixed, thermic Typic Udipsamments	excessively drained
Menfro	Fine-silty, mixed, superactive, mesic Typic Hapludalfs	well drained
Portageville	Fine, smectitic, calcareous, thermic Vertic Endoaquolls	poorly drained
Reelfoot	Fine-silty, mixed, superactive, thermic Aquic Argiudolls	somewhat poorly drained
Rueter	Loamy, mixed, superactive, mesic, shallow Vitritorrandic Haploxerolls	well drained
Sharkey	Very-fine, smectitic, thermic Chromic Epiaquerts	poorly drained
Taumsauk	Loamy-skeletal, mixed, active, mesic Lithic Hapludults	somewhat excessively drained
Tiptonville	Fine-silty, mixed, superactive, thermic Oxyaquic Argiudolls	moderately well drained
Wilbur	Coarse-silty, mixed, superactive, mesic Fluvaquentic Eutrudepts	moderately well drained

Laboratory Protocols

An aqua regia digestion was employed to obtain a near total estimation of elemental abundance associated with all but the most recalcitrant soil chemical environments (Aide and Fasnacht, 2010). Aqua-regia digestion does not appreciably degrade quartz, albite, orthoclase, anatase, barite, monazite, sphene, chromite, ilmenite, rutile and cassiterite; however, anorthite and phyllosilicates are partially digested. Homogenized samples (0.75g) were equilibrated with 0.01 liter of aqua-regia (3 mole nitric acid: 1 mole hydrochloric acid) in a 35ºC incubator for 24 hours. Samples were shaken, centrifuged and filtered (0.45 μm), with a known aliquot volume analyzed using inductively coupled plasma – mass spectrometry (ICP-MS). The aqua regia digestion procedure was performed by Activation Laboratories (Toronto, Canada). In this procedure, selected samples were duplicated and known reference materials were employed to guarantee analytical accuracy. The detection limit for Fe was 0.01 percent, the detection limit for Th was 0.1 mg/kg and the U detection limit was 0.1 mg/kg.

A hot water extraction was performed to recover only the most labile or potentially labile fractions. A hot water extraction involved equilibrating 0.5 g samples in 0.02 L distilled-deionized water at 80ºC for one hour followed by 0.45μm filtering and elemental determination using ICP-MS. For the water extraction, selected samples were duplicated and reference materials were employed to guarantee analytical precession. The detection limits for Th and U were 0.1 μg/kg-soil.

Some soil pedons were grouped for statistical analysis because they represent the same soil series and were located in close proximity. Regression analysis and Pearson correlations were performed using Microsoft Excel. To test the null hypothesis that the linear regression slope ($\beta 1$) is zero, the variance of the error about the slope was estimated to create confidence intervals ($\alpha = 0.05$) were developed (Johnson and Kuby, 2004). Paired T-tests were used to establish correlation confidence.

RESULTS AND DISCUSSION

Relationship Involving Iron, Thorium and Uranium and Soil Profile Depth

The thorium and uranium soil profile distributions for 25 soil series show a range in distribution trends.

Table 3. Parent materials and vegetation class

Soil Series	Parent Materials	Vegetation
Alred	loess over cherty limestone residuum	mixed deciduous forest
Amagon	fine-silty alluvium	mixed deciduous forest
Broseley	sandy alluvium	row crop
Calhoun	fine-silty alluvium	mixed deciduous forest
Caneyville	loess	mixed deciduous forest
Clana	coarse-textured terrace (alluvium)	irrigated row crop
Commerce	loamy alluvial sediments	row crop
Crowley	fine-silty alluvium	irrigated row crop
Dubbs	silt over loamy sand alluvium	row crop
Foley	silty terrace material high in sodium	irrigated row crop
Frenchmill	fine-silty colluvium of acid igneous rocks	mixed deciduous forest
Heldebrecht	loess over dolomite residuum	mixed deciduous forest
Irondale	loess over rhyolite residuum and colluvium	mixed deciduous forest
Killarney	slope alluvium mixed with loess and slope alluvium	mixed deciduous forest
Knobtop	loess over rhyolite	mixed deciduous forest
Lilbourn	coarse-textured terrace (alluvium)	irrigated row crop
Malden	coarse-textured terrace (alluvium)	irrigated row crop
Menfro	loess	mixed deciduous forest
Portageville	fine-textured alluvium	irrigated row crop
Reelfoot	silty alluvium	row crop
Rueter	loess over cherty limestone residuum	mixed deciduous forest
Sharkey	fine textured alluvium	irrigated row crop
Taumsauk	colluvium or residuum weathered from rhyolite	mixed deciduous forest
Tiptonville	silty-textured alluvium	irrigated row crop
Wilbur	silty-textured alluvium	irrigated row crop

Table 4. Whole soil Mean Iron, Thorium, and Uranium concentrations (mg/kg-soil)

	----------------Iron--------------		--------------Thorium---------		-----------Uranium----------	
Soil Series	**Mean**	**Standard Deviation**	**Mean**	**Standard Deviation**	**Mean**	**Standard Deviation**
Alred	24,430	12,600	9.20	5.82	0.66	0.20
Amagon	16,042	5,490	3.51	0.82	1.28	0.26
Broseley	10,530	3,940	2.30	1.33	0.63	0.13
Calhoun	17,380	7,660	3.91	1.82	1.08	0.17
Caneyville	22,150	7,340	5.39	2.76	1.01	0.39
Clana	12,030	3,320	2.41	1.28	0.84	0.21
Commerce	18,190	3,940	5.82	0.42	1.31	0.24
Crowley	25,190	9,520	7.62	1.78	1.98	0.93
Dubbs	15,690	4,420	4.16	1.63	1.00	0.12
Foley	28,210	4,120	8.17	1.27	1.69	0.23
Frenchmill	20,740	6,480	5.51	0.75	1.36	0.36
Hildebrecht	14,680	5,920	6.69	3.82	1.15	0.38
Irondale	21,580	5,470	6.77	2.78	1.30	0.30
Killarney	13,000	1,920	3.02	0.71	0.92	0.19
Knobtop	22,180	6,100	4.30	1.29	1.42	0.36
Lilbourn	13,090	1,350	2.71	1.02	0.88	0.17
Malden	10,590	1,190	2.42	0.50	0.73	0.11
Menfro	18,270	6,850	5.73	2.78	1.12	0.30
Portageville	37,360	2,710	10.33	0.73	2.66	0.13
Reelfoot	18,910	3,900	5.53	1.61	1.03	0.05
Rueter	16,700	6,660	5.74	2.71	0.58	0.13
Sharkey	30,330	6,940	8.58	1.90	2.80	1.30
Taumsauk	26,640	6270	7.72	1.97	1.36	0.30
Tiptonville	15,450	6,230	4.23	1.97	1.05	0.10
Wilbur	17,720	2,730	4.16	1.08	1.00	0.16

All "pooled" pedons were adjacent, sharing a similar soil classification and having identical land use.

Table 3 displays the Fe, Th and U mean and standard deviation among the soil series selected for this study. The mean whole soil Th concentrations range from 2.3 to 10.3 mg Th/kg, whereas the mean whole soil U concentrations range from 0.58 to 2.80 mg U/kg. The mean whole soil Fe concentrations range from 10,530 to 37,360 mg Fe/kg. Coarse-textured soils (Broseley, Clana, Lilbourn and Malden) had the smallest concentrations of Th and U, with the exception that the loamy-textured Rueter pedon had a mean thorium concentration greater than mean thorium concentration of all of the sampled pedons and also exhibited the smallest mean U concentration of the study area.

Thorium correlations with Fe (Table 5) demonstrate appreciable correspondence for most of the sampled pedons. Four pedons did not have significant correlations involving Th with Fe (Commerce, Killarney, Portageville and Wilbur) with all four of those pedons having correspondingly small Fe standard deviations, suggesting that the lack of Fe variation contributed to the lack of Fe-Th correlation.

Sampled soils (pedons) having argillic horizons displayed greater Fe, Th and U concentrations in the argillic horizons than the overlying eluvial horizons. An example is the Th concentrations in the argillic horizons of the Menfro pedons (Figure 3).

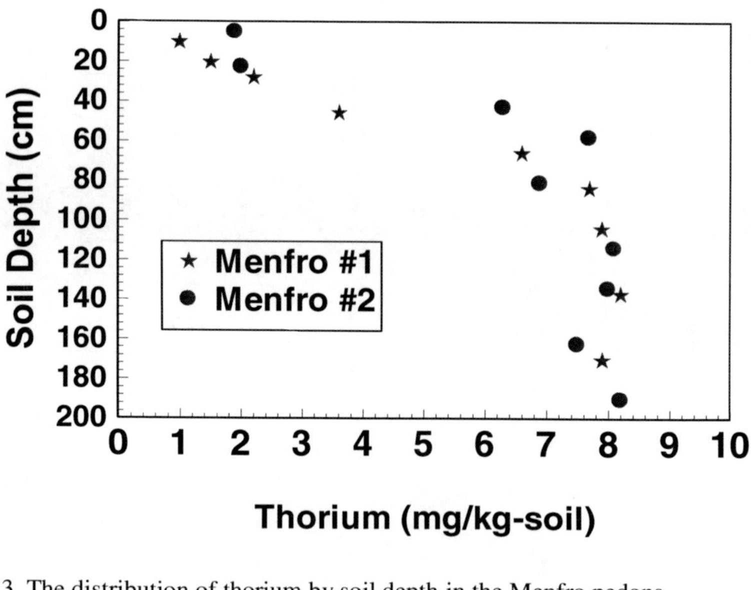

Figure 3. The distribution of thorium by soil depth in the Menfro pedons.

Table 5. Statistical evaluation of iron, thorium, and uranium concentrations in selected soil profiles

	Th / U		Correlation (r)	
Soil Series	Mean	Standard Deviation	Th-U	Th-Fe
Alred	12.6	6.1	0.96	0.98
Amagon	2.7	0.6	0.73	0.88
Broseley	3.5	1.4	0.84	0.91
Calhoun	3.6	1.3	0.75	0.35
Caneyville	5.1	1.0	0.96	0.84
Clana	2.8	0.9	0.85	0.95
Commerce	4.6	0.9	-----	-----
Crowley	4.0	1.0	0.64	0.81
Dubbs	4.1	1.5	-----	0.96
Foley	4.8	0.4	0.88	0.87
Frenchmill	3.8	0.3	0.95	0.89
Hildebrecht	5.8	1.5	0.81	0.99
Irondale	5.0	1.7	0.93	0.96
Killarney	3.9	1.8	-----	-----
Knobtop	3.0	0.3	0.98	0.94
Lilbourn	3.3	1.3	-----	0.40
Malden	3.4	0.7	0.37	0.69
Menfro	4.8	1.6	0.96	0.98
Portageville	3.9	0.4	-----	-----
Reelfoot	5.4	1.6	-----	0.97
Rueter	9.4	2.8	0.99	0.98
Sharkey	3.4	1.2	0.35	0.54
Taumsauk	1.4	0.4	0.97	0.97
Tiptonville	4.1	2.0	-----	0.71
Wilbur	4.3	1.5	-----	-----

All listed values are significantly valid at least to $\alpha = 0.05$.

The Crowley pedon was separated into its sand, silt and clay separates because the sand separate exhibited nodules of Fe- and Mn- oxyhydroxides. The sand and clay separates possess greater Fe, Th and U concentrations than the silt separate and corresponding whole soil (Figures 4 and 5). The clay, silt and to some extent the sand separate maintained reasonably consistent Th/Fe and U/Fe ratios, inferring that pedogenic processes responsible for accumulation of Th and U among the separates did not dramatically exhibit any positive or negative preference for Th with respect to U. The Foley pedons similarly exhibited pedogenic Fe- and Mn-bearing nodules in the sand separate. The sand separate demonstrated greater Fe, Th and U in the sand separates (Figures 6 and 7). The pooled Frenchmill, Irondale, Killarney and Knobtop were pooled because they formed in residuum of rhyolite. The clay separates were isolated from selected soil horizons to determine their Fe, Th and U contents. Figures 8 and 9 demonstrate that the clay separate typically shows greater Fe, Th and U concentrations and that little Th or U preference in its association with Fe is observed between the whole soil and the clay separate.

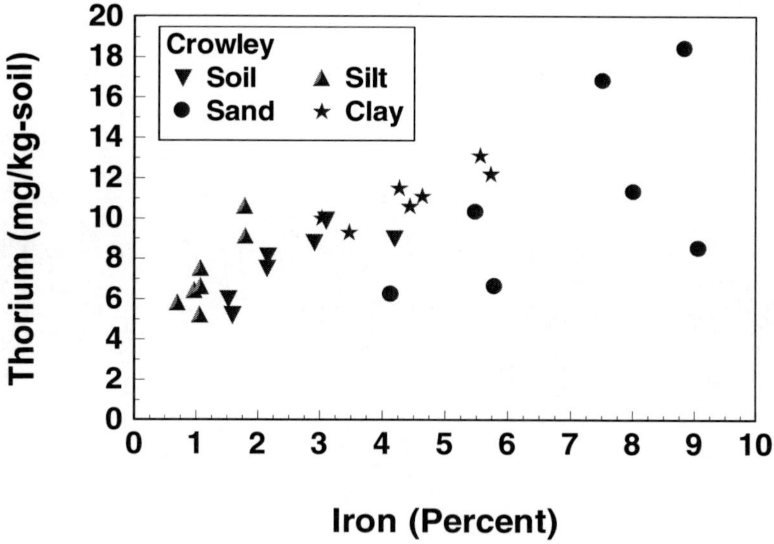

Figure 4. The thorium concentrations as a function of the corresponding Fe concentrations for the Crowley pedon.

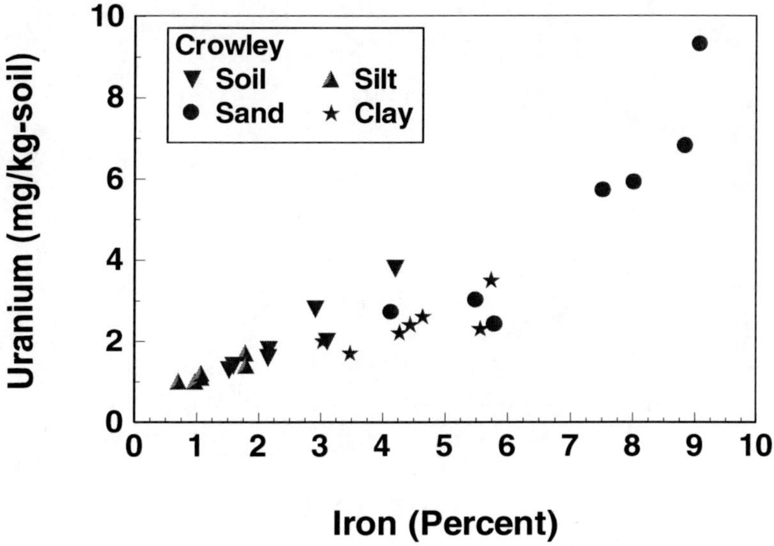

Figure 5. The uranium concentrations as a function of the corresponding Fe concentrations for the Crowley pedon.

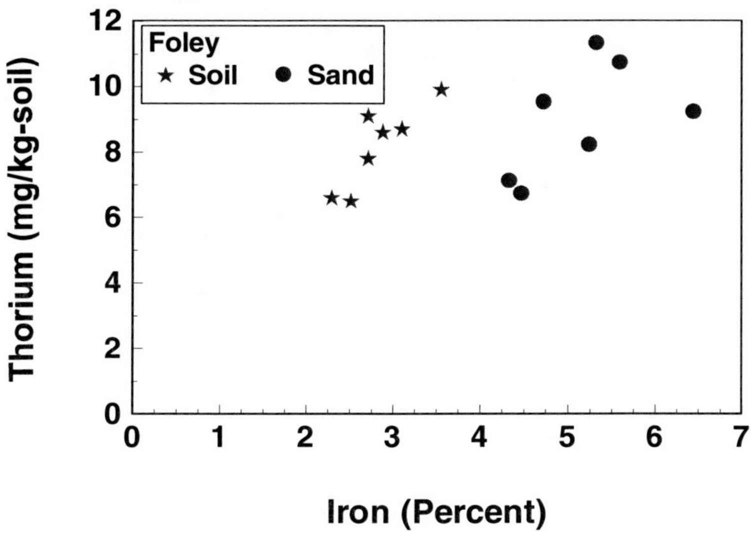

Figure 6. The thorium concentrations as a function of the corresponding Fe concentrations for the Foley pedon.

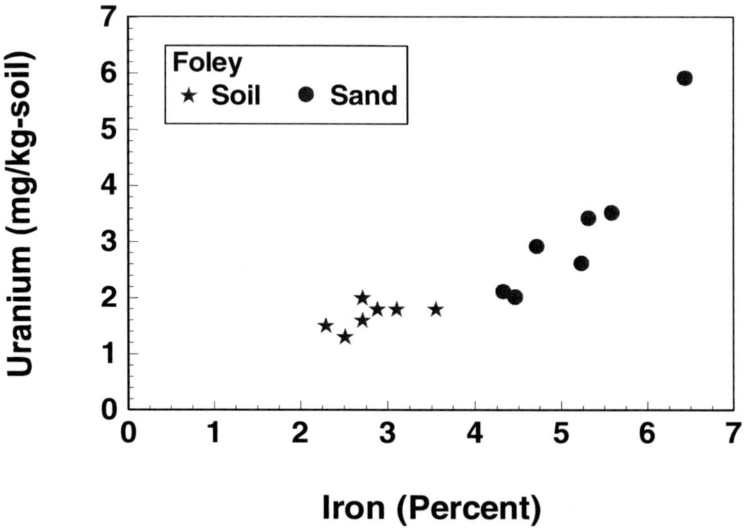

Figure 7. The uranium concentrations as a function of the corresponding Fe concentrations for the Foley pedon.

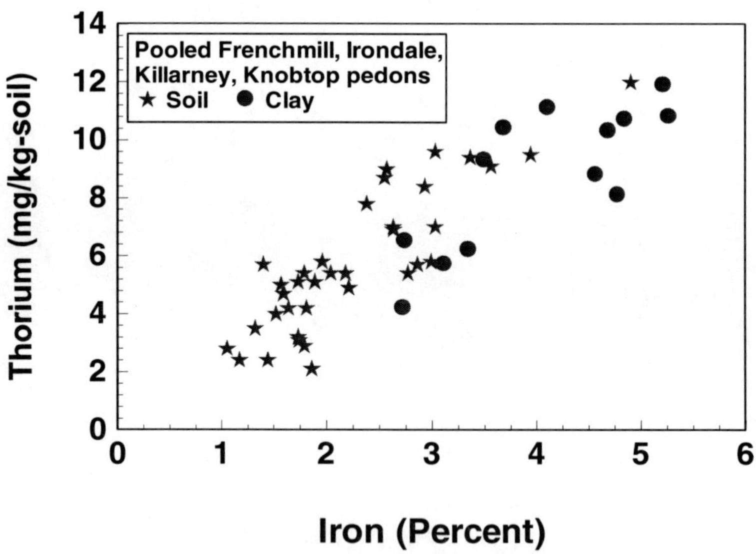

Figure 8. The thorium concentrations as a function of the corresponding Fe concentrations for the pooled Frenchmill, Irondale, Killarney and Knobtop pedons.

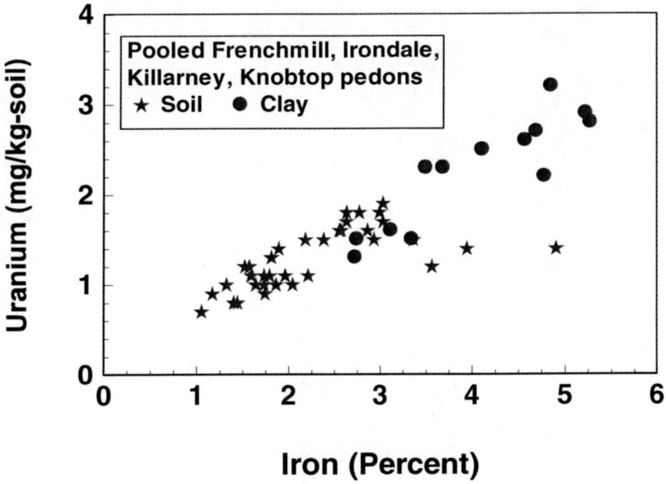

Figure 9. The uranium concentrations as a function of the corresponding Fe concentrations for the pooled Frenchmill, Irondale, Killarney and Knobtop pedons.

The Menfro pedons are developed entirely in deep loess. Thorium and Uranium are closely associated with Fe, with the argillic horizons having greater Fe expression than the eluvial horizons (Figures 10, 11).

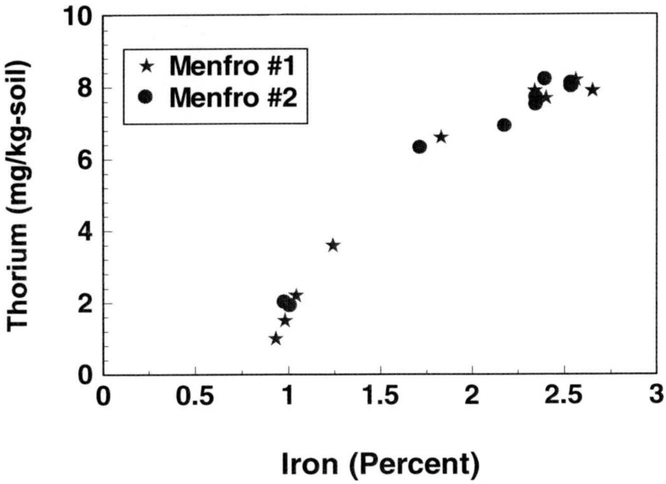

Figure 10. The thorium concentrations as a function of the corresponding Fe concentrations for the Menfro pedons.

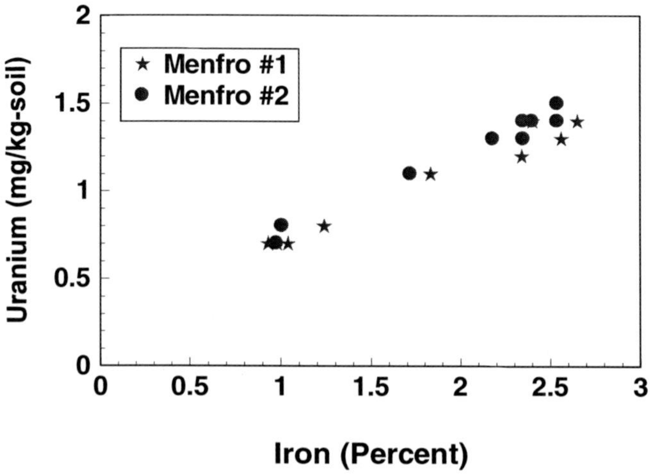

Figure 11. The uranium concentrations as a function of the corresponding Fe concentrations for the Menfro pedons.

Water Extracts Involving Zinc and Cadmium

Water extractions of selected soil horizons represents metal recovery of the most immediately plant available fraction within the soil environment. Water extractable Th and U were assessed for the Amagon, Calhoun, Sharkey, Caneyville, Wilbur Clana, Malden, and Lilbourn pedons. As expected, the relationship between the Th and U aqua regia digestion values and their corresponding Th and U water extraction values has no correlation. In general, the mean U water extraction values are greater than the mean Th water extraction values (Figures 12 and 13). The Amagon pedons had the greatest mean Th water extraction concentrations, whereas the Calhoun pedon had the greatest mean U water extraction concentrations (9 μg Th/kg for the Amagon pedons and 16 μg U/kg for the Calhoun pedon). Conversely, the Caneyville and Wilbur pedons exhibited the smallest Th and U water extraction recoveries.

In general, the eluvial horizons demonstrated greater U and Th water extraction values (Figure 14 and 15), inferring that organic complexes of Th and U are important fractions for augmenting Th and U in the water fraction.

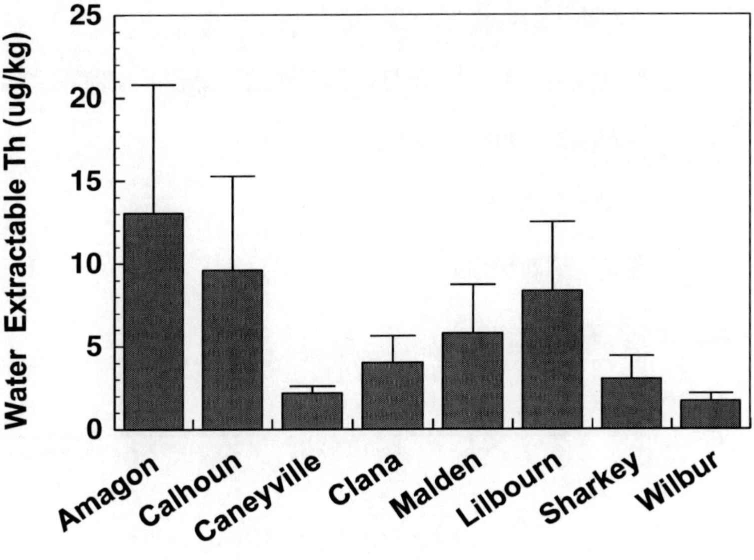

Figure 12. Mean water extractable thorium for selected soils. Error bars are standard deviations.

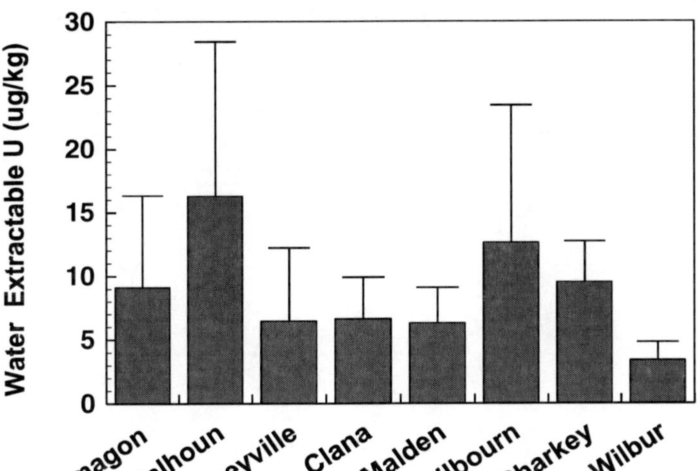

Figure 13. Mean water extractable uranium for selected soils. Error bars are standard deviations.

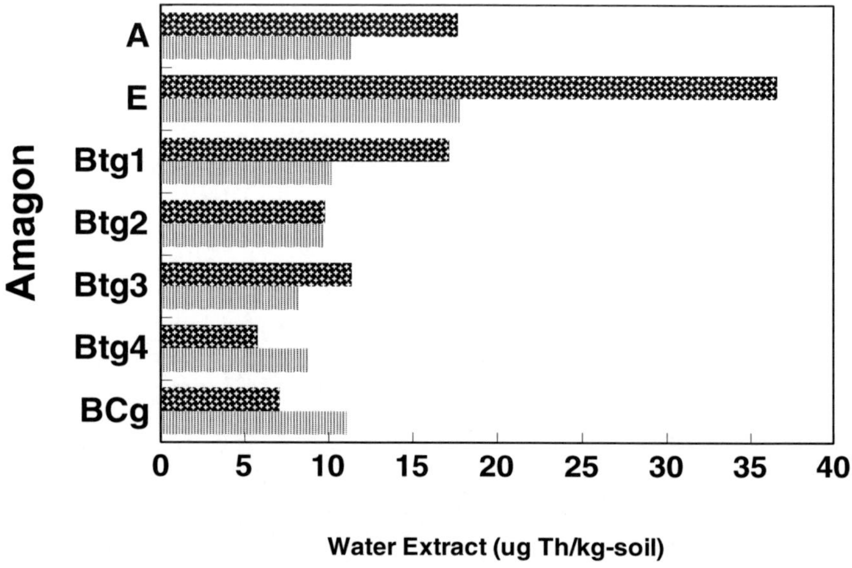

Figure 14. Water extractable thorium for horizons of two pedons of the Amagon series.

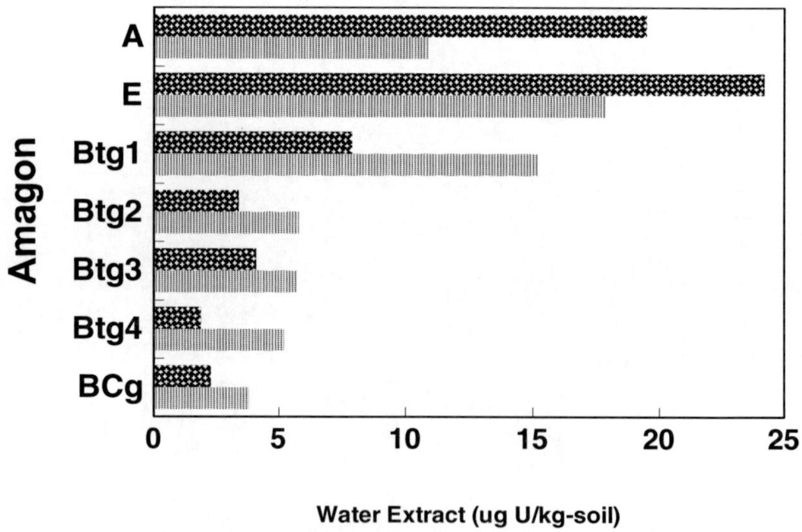

Figure 15. Water extractable uranium for horizons of two pedons of the Amagon series.

THORIUM AND URANIUM SOIL PROFILE

Distributions and Soil Morphology

The soil profile Th and U concentration distributions fall generally into two soil categories based on the presence of diagnostic soil horizons: the argillic soil horizon and the cambic soil horizon. The presence of Fe-oxyhydroxides associated with phyllosilicates provides greater site-bonding for Th and U. The presence of argillic horizons on stable landscapes will likely correlate with Th and U soil profile maxima given the abundance of binding sites. The clay eluviation-illuviation process likely provides interhorizon transport of Th and U. The soil organic matter component is not a dominant Th and U fraction; however, organic complexation of Th and U may promote plant uptake given the relative bonding strength of Fe-oxyhydroxides and the ineretness of U and Th co-precipitation with the Fe-oxyhydroxide fraction.

REFERENCES

Aide, M.T., K. Whitener, E. Westhoff and J. Kelly. 2008. Effectiveness of triple superphosphate amendments in alleviating soil lead accumulation in Missouri Alfisols. *Soil Sediment Contamination* 17:630-642.

Aide, M.T. 2009. Three Missouri Alfisols impacted by aeolian deposition of Pb-Zn-Cd-Ag bearing mine tailings. *Soil Sediment Contamination.* 18:43-54.

Aide, Michael and I. Braden. 2009. Inter-horizon transport of elements in Ultisols developed in a loess mantle mixed with rhyolite residuum in the St. Francois Mountains of Missouri. *Soil Sci.* 174:210-219.

Aide, M.T., and M. Fasnacht. 2010. Estimating trace element availability in soils having a seasonal water table using commercially available protocols. *Comm. Soil Sci. and Plant Anal.* 41:1159-1177. (doi: 10:1080/00103621003721379)

Aide, Michael, Indi Braden, Neil Hermann, David Mauk, Wesley Mueller, Sven Svenson, and Julie Weathers. 2011. Using Lanthanum, neodymium and thorium as conservative indexing elements to indicate soil lead deposition. *Soil and Soil Contamination*. 22:132-150.

Allison, J.D., D.S. Brown, and K.L. Novo-Gradac. 1991. *Minteqa2/Prodefa2, A geochemical assessment model for environmental systems: version 3.0.*

Environmental Research Laboratory, Office of Research and Development, USEPA, Athens, GA.

Baes, C.F., and R.E. Mesmer. 1976. *The hydrolysis of cations.* Wiley-Interscience. N.Y.

Barnett, M.O., P.M. Jardine, S.C. Brooks, and H.M. Selim. 2000. Adsorption and transport of uranium (VI) in subsurface media. *Soil Sci. Soc. Am. J.* 64:908-917.

Barnett, M.O., P.M. Jardine, and S.C. Brooks. 2002. Uranium (VI) adsorption to heterogeneous subsurface media: Application of a surface complexation model. *Environ. Sci. Technol.* 36:937-942.

Bednar, A.J., D.B. Gent, J.R. Gilmore, T.C. Sturgis, and S.L. Larson. 2004. Mechanisms of thorium migration in a semiarid soil. *J. Environ. Qual.* 33:2070-2077.

Bidoglio, G., A. De Plano, and L. Righetto 1989. Interaction and transport of plutonium-humic acid particles in groundwater environments. *Mat. Res. Soc. Symp. Proc.* 127:823-830.

Carter, M.R. 1993. Soil sampling and methods of analysis. *Canadian Soc. Soil Sci.* Ottawa, Ontario, Canada.

Chen, J.P., and S. Yiacoumi. 2002. Modeling of depleted uranium transport in subsurface systems. *Water, Air and Soil Polution.* 140:173-201.

Chopping, P.M., and G.R. Shanbhag. 1981. Binding of uranyl to humic acid. *J. Inorg. Nucl. Chem.* 43:3369-3374.

Cygan, R.T. 2002. Molecular models of radionuclide interaction with soil minerals. *In* (Peng-Chu Zhang and Patrick V. Brady, eds,) Geochemistry of soil radionuclides. *Soil Science Society America Special Publication* 59. Soil Science Society America, Madison, WI.

Davis, J.A., T.E. Payne, and T.D. Waite. 2002. Simulating the pH and pCO2 dependence of uranium(VI) adsorption by a weathered schist with surface complexation models. *In* (Peng-Chu Zhang and Patrick V. Brady, eds,) Geochemistry of soil radionuclides. Soil Science Society America Special Publication 59. *Soil Science Society America*, Madison, WI.

Dowdall, M, J.P. Gwynn, G.W. Gabrielsen, and B. Lind. 2005. Assessment of elevated radionuclide levels in soils associated with an avian colony in a high arctic environment. *Soil Sediment Contamination* 14:1-11.

Ebbs, S.D., D.J. Brady, and L.V. Kochian. 1998. Role of uranium speciation in the uptake and translocation of uranium by plants. *J. Experimental Botany* 49:1183-1190.

Essington, M. 2004. *Soil and water chemistry: An integrative approach.* CRC Press, Boca Raton, FL.

Greenwood, N.N., and A. Earnshaw. 1984. *Chemistry of the elements.* Pergamon Press, New York.

Hamby, D.M., and A.K. Tynybekov. 2002. Uranium, thorium, and potassium in soils along the shore of Lake Issyk-Kyol in the Kyrghyz Republic. *Environ. Monitoring and Assessment* 73:101-108.

Hichung, M. 1989. Equilibrium ultrafiltration of hydrolyzed thorium(IV) solutions. *Bulletin Korean Chemical Soc.* 10:270-272.

Honeyman, B.D., and J. F. Ranville. 2002. Colloid properties and their effects on radionuclide transport through soils and groundwaters. *In* (Peng-Chu Zhang and Patrick V. Brady, eds,) Geochemistry of soil radionuclides. Soil Science Society America Special Publication 59. *Soil Science Society America*, Madison, WI.

Hsi, C-K.D., and D. Langmuir. 1985. Adsorption of uranyl onto ferric oxyhydroxides: Application of the surface complexation site binding model. *Geochim Cosmochim Acta* 49:2423-2432.

Johnson, R., and P. Kuby. 2004. *Elementary Statistics.* Brooks Cole Publ., Belmont, CA

Johnson, W.H., B.J. Buck, H. Brogonia, and A.L. Brock. 2004. Variations in depleted uranium sorption and solubility with depth in arid soils. *Soil Sediment Contamination.* 13:533-544.

Kaplan, D.I., and S.M. Serkiz. 2001. Quantification of thorium and uranium sorption to contaminated sediments. *J. Radioanal. Nucl. Chem.* 248:529-535.

Kumar, A., R.K. Singhal, J. Preetha, K. Rupali, U. Narayanan, S. Suresh, M.K. Mishra and A.K. Ranade. 2008. Impact of tropical ecosystem on the migrational behavior of K-40, Cs-137, Th-232 and U-238 in perennial plants. *Water, Air Soil Pollution* 192:293-302.

Langmuir, D. and J.S. Herman. 1980. The mobility of thorium in natural waters at low temperatures. Geochim. Cosmochim Acta 44:1753-1766.

Langmuir, D. 1997. *Aqueous environmental geochemistry.* Prentiss-Hall, Upper Saddle River, New Jersey.

Martinez-Aguirre, A., and R. Perianez. 2001. Sedimentary speciation of U and Th isotopes in a marsh areaat the southwest of Spain. *J. Radioanal. Nucl. Chem.* 247:45-52.

McKinley, J.P., J.M. Zachara, S.C. Smith, and G.D. Turner. 1995. The influence of uranyl hydrolysis and multiple site-binding reactions on the adsorption of U (VI) to montmorillonite. *Clays Clay Miner.* 43:586-598.

Moon, J.W., Y. Roh, T.J. Phelps, D.H. Phillips, D.B. Watson, Y.J. Kim, and S.C. Brooks. 2006. Physiochemical and mineralogical characterization of soil-saprolite cores from a research site, Tennessee. *J. Environ. Qual.* 35:1731-1741.

Morton, L.S., C.V. Evans, and G.O. Estes. 2002. Natural uranium and thorium distributions in podzolized soils and native blueberry. *J. Environ. Qual.* 31:155-162.

Murphy, R.J., J.J. Lenhart, and B.D. Honeyman. 1999. The sorption of thorium(IV) and uranium(VI) to hematite in the presence of natural organic matter. *Colloids Surf. A.* 157:47-62.

Norvell, W.A. 1991. Reactions of metal chelates in soils and nutrient solutions. *In* Mortvelt, J.J., F.R. Cox, L.M. Shuman, and R.M. Welch (eds). p.187-227. Micronutrients in agriculture. *Soil Sci. Soc. Am. 2nd ed.* - SSSA Book Series, no. 4.

Phillips, D.H., D.B. Watson, Y. Roh, T.L. Mehlhorn, J.W. Moon, and P.M. Jardine. 2006. Distribution of uranium in weathered fractured saprolite/shale and ground water. *J. Environ. Qual.* 35:1715-1730.

Raju, K.K., and A.N. Raju. 2000. Biogeochemical investigation in south eastern Andhra Pradesh: the distribution of rare earths, thorium and uranium in plants and soils. *Environ. Geol.* 39:1102-1106.

Riese, A.C. 1982. *Adsorption of radium and thorium onto quartz and kaolinite: A comparison of solution/surface equilibria models.* Ph.D. thesis. T-2625 Colorado School Mines, Golden, CO.

Roh, Y., S.R. Lee, S.K. Choi, M.P. Elless, and S.Y. Lee. 2000. Physiochemical and mineralogical characterization of uranium-contaminated soils. *Soil Sediment Contamination* 9:463-486.

Runde, W. 2002. Geochemical interactions of actinides in the environment. *In* (Peng-Chu Zhang and Patrick V. Brady, eds,) Geochemistry of soil radionuclides. *Soil Science Society America Special Publication* 59. Soil Science Society America, Madison, WI.

Runde, W., M.P. Neu, S.D. Conradson, D.Li, M. Lin, D.M. Smith, C.E. Van-Pelt, and Y. Xu. 2002. Geochemical speciation of radionuclides in soil and solution. *In* (Peng-Chu Zhang and Patrick V. Brady, eds,) Geochemistry of soil radionuclides. *Soil Science Society America Special Publication 59.* Soil Science Society America, Madison, WI.

Santos, Jr., J.A., R.S. Amaral, C.M. Silva, R.S.C. Menezes, and J.D. Bezerra. 2009. Radium-228 as an indicator of thorium-232 presence in a soil in Pernambuco, Brazil. Bull. *Environ. Contam. Toxicol.* 82:650-652.

Sato, T., T. Murakami, N. Yanase, H. Isobe, T.E. Payne, and P. Airy. 1997. Iron nodules scavenging uranium from ground water. *Environ. Sci. Technol.* 31:2854-2858.

Shahandeh, H., and L.R. Hossner. 2002. Role of soil properties in phytoaccumulation of uranium. *Water, Air, and Soil Pollution* 141:165-180.

Shtangeeva, I., and S. Ayrault. 2004. Phytoextraction of thorium from soil and water media. *Water Air Soil Pollution* 154:19-35.

Soil Survey Division Staff. 1993. *Soil survey manual.* USDA Handb. 18. U.S. Gov. Print. Office. Washington, DC.

Stewart, B.D., J. Neiss, and S. Fendorf. 2007. Quantifying constraints imposed by calcium and iron on bacterial reduction of uranium(VI). *J. Environ. Qual.* 36:363-372.

Turner, G.D., J.M. Zachara, J.P. McKinley, and S.C. Smith. 1996. Surface charge properties and UO_2^{2+} adsorption on a subsurface smectite. *Geochim. Cosmochim Acta* 60:3399-3414.

Varinlioglu, A., and A. Kose. 2005. Determination of natural and artificial radionuclide levels in soils of western and southern coastal area of Turkey. *Water, Air, Soil Pollution.* 164:401-407.

Waite, T.O., J.A. Davis, T.E. Payne, G.A. Waychunas and N. Xu. Uranium (VI) adsorption to ferrihydrate: Application of a surface complexation model. *Geochim Cosmochim Acta* 58:5465-5478.

Wanty, R.B., and D.K. Nordstrom. 1995. Natural radionuclides. In (W.M. Alley, ed) *Regional ground-water quality.* Van Nostrand Reinhold. New York.

In: Thorium
Editor: Renae Jamison
ISBN: 978-1-63321-309-8

Chapter 3

INFLUENCE OF A COAL-FIRED POWER PLANT ON THE THORIUM LEVELS IN SOILS AND ON THE RADIOACTIVE HAZARD FOR THE POPULATION

Elena Charro[1,2,*] ***and Rafael Pardo***[3]
[1]LIBRA Laboratory, Universidad de Valladolid, España
[2]Faculty of Education, Universidad de Valladolid, Spain
[3]Faculty of Science, Universidad de Valladolid, Spain

ABSTRACT

Coal contains trace amounts of natural radionuclides, such as ^{232}Th. In the burning process in a coal-fired power-plant (CFPP), ashes become enriched in this radionuclide, which can be released as fly ash from the CFPP and deposed from the atmosphere on the nearby top soils, modifying the natural radioactivity background levels, and subsequently increasing the total radioactive dose received by the population. Thus, the estimation of thorium levels in the soils can be used as a tool to know the radiological impact of a CFPP. In this paper, the radiological impact of a CFPP in Spain has been studied. The activities of Thorium and other radionuclides have been measured by gamma spectrometry in coal, ash, sediment, waste and soil samples. From these results, the distribution of Thorium in soils and if deposition and migration has been taken place can

* Corresponding author. Tel: 00 34 983 423466. E-mail: echarro@dce.uva.es.

be analysed taking into account its physical-chemistry properties. Analysis of Variance (ANOVA), Principal Component Analysis (PCA), Hierarchical Clustering Analysis (HCA), and kriging mapping have been applied to the experimental dataset. Finally, the mean annual outdoor effective dose has been evaluated as 71 μSv, which is very close to the average UNSCEAR value of 70 μSv, and this means that Th concentrations found have not influence on the radioactivity levels measured in the studied area, and finally, thus confirming that the presence of the CFPP has no radioactive impact in the population.

Keywords: Thorium, radionuclides; coal fired power plant; coal, soils; statistical analysis; dose assessment

1. INTRODUCTION

Thorium is surprisingly common in the Earth's crust, being almost as abundant as lead and three times more abundant than uranium. It is found in small amounts in most rocks and soils. The determination of Uranium and thorium has been used to date hominid fossils, but one of the most important applications is its use, like uranium, as nuclear fuel. Thorium can be burnt in a nuclear reactor, without generating plutonium. There is probably more untapped energy available for use in the thorium present in the minerals of the earth's crust, than in the combination of uranium and fossil fuel sources. Indeed, much of the internal heat of the earth has been attributed to thorium and uranium.

Until the inherent dangers associated with its radioactivity were realized, thorium and its compounds found some important retail uses such gas mantles and toothpastes. Thorium is also present in consumer products such as ceramic glazes, lantern mantles and welding rods. Although we cannot expect environmental problems from these thorium-containing products (because of their small size), large thorium releases may be harmful to plants and animals. Because thorium oxide is highly insoluble, only small amounts of this element circulates through the environment. However, since thorium is always naturally present in the environment, people are exposed to tiny amounts in air, food and water, which usually pose little health hazard. Thorium waste disposal must be in accordance with appropriate federal, state, and local regulations. The thorium-containing products, if unaltered by use, may be disposed of by their treatment at a solid waste permitted facility or as advised by the local hazardous waste regulatory authority. The principal concern from low to moderate level exposure to ionizing radiation is an increased risk of cancer. Studies have shown that thorium dust inhalation causes an increased risk of developing lung and pancreas cancer. Bone cancer risk also increases because thorium may be stored in bones.

Coal contains natural radioactive elements in similar concentrations that other similar sedimentary rocks [1]. Lignite, or brown coal, is sometimes an exception, having a higher-than-average content of radionuclides, particularly Uranium and Thorium. Coal mining, washing, storing or combustion processes can redistribute these radionuclides, causing local increases over the natural background radioactivity levels. The radiological impact of coal fired power plants (CFPP) has been one of the most important environmental problems under study in the last years. Many authors [2-8] focused their works on the use of the brown coal, finding a real CFPP impact on surrounding areas, since the amount of radioactivity discharged by CFPPs could be two orders of magnitude higher than that done by plants using ordinary coal [4]. Local effects of coal mining and other processes of coal industrial cycle have been less studied. Although CFPP ashes could be considered a major source of pollution due to coal industry, there are other possible pathways for contamination by radionuclides associated to the industrial cycle of coal. Possible environmental pathways of radionuclides have been highlighted, starting at coal mining, which could produce radionuclide pollution in drainage waters, and paying after attention to coal washing and CFPP operations.

After extraction, about 50% of the rock of mine (ROM) coal is washed. During this physical coal cleaning, the elements become partitioned among cleaned coal, coal wastes and the washing water [3]. Mine drainage, represents an environmental threat and usually has presented high concentrations of heavy metals as well as some radioelements [9]. Cleaning wastes piles are located normally in public areas, where are affected by meteorological conditions, so leaching of radionuclides from piles to soil and streams is thus possible. Furthermore, the beds of nearby streams and rivers might have been polluted by past mining activities, when waste waters were directly released to the environment. The study of radionuclide contents in sediments of an area under study, will provided very useful information about the real impact of coal industry. Radionuclides concentrations in sediments, usually related with the concentrations in nearby soils, could be affected by the radionuclides leached from waste piles. However, a large number of parameters affect to the distribution of radionuclides in sediments, thus conditioning their mobility and interferences with other compartments, mainly water and suspended material.

Washed coal is then burned in the boiler of CFPPs, and all the natural radionuclides of the coal remain among the incombustible elements. These non-volatile combustion products originate ash particles, some of which fall out in the boiler as “bottom ash”. The remaining particles, commonly named “fly-ash”, are carried out by the exhaust gas stream toward the emission control equipment. The non-collected ash is emitted to the atmosphere with the flue gas. Therefore, ash releases should be considered a source of radiation to the population living in the vicinity of a CFPP (and also a potential hazard to the CFPP personnel), through the inhalation of suspended ash material, or the ingestion of foodstuffs after deposition on crops and pasture, or by external irradiation due to the deposited activity [4, 10, 11]. Fly ash is a waste product used in several countries as a component of building materials, and also as a potential nutrient supplement for degraded agricultural soils [12], although a previous and exhaustive study of their heavy metal and radionuclide contents would be necessary.

The goal of this chapter has been the analysis of the potential environmental contamination by Thorium, released along other radionuclides to environment, from different processes of the coal industrial cycle. For this purpose, a radiological study in a mountain region in the north of Palencia (Spain), affected by coal mining activities and a CFPP, has been performed. Activity concentrations of ^{232}Th, and other radionuclides were determined in coal, ash, waste, sediment and soil samples. The main emphasis of this study has been made on the analysis of the distribution and mobility of thorium in the above mentioned samples, taking into account how this radionuclide can be involved in future environmental problems in the studied area. The impact of this radionuclide is also compared with the contribution of the rest of radionuclides found in the coal.

2. Methodology

2.1. Study Area

The area is a mountain region about 100 km^2, located at the central-north of Spain, in the Palencia province. The weather is continental with temperatures varying between 4ºC in winter and 14ºC in summer, with a median rainfall of 1000 mm (that becomes snow in winter. The soil use is for animal pasture, and coal mining is the main industrial activity in the area. The geology of the area is marked by calcareous mountain and the presence of coal materials. The local mining activities started at the beginning of last century, reaching their highest productivity during the late sixties. Nowadays, the coal production in the region is about half million tons per year, extracted by both underground and surface mining. The power plant of Velilla del Río Carrión (42°49’ N, 4°51’ W), is placed next to the Carrión river and was built in 1964 (a new second generator was added in 1984 with a new higher stack). It has a storage capacity of 1 million tons of coal, and consumes on average 800.000 tons per year.

2.2. Sample Collection and Measurement

Figure 1 shows a topographical scheme of the studied with altitude lines and the CFPP, the villages of Guardo and Velilla, the Carrión River and the main roads highlighted. Our study was focused on five types of samples: coal, ash, waste material, sediment and soil samples. Coal and ash samples were provided by the CFPP. Four samples of coal were collected from an old storage area and the ash was obtained from the ash repository (where ashes collected by electrostatic filters are located) before releasing. The samples of waste materials were taken from the surface and the top of the waste pile. Two different groups of piles were considered. One group (“old wastes” OW-1, OW-2, and OW-3) was located at an old mining area with no recent activity, and where no wastes were deposited since years ago. The other group of waste materials (“new wastes” NW-1 and NW-2) were collected in an active mining area, where more wastes are been recently added. All these points can be located in Figure 1. Nine sediment samples (SD-1 to SD-9) and two reference soil samples (S-1 and S-2) were collected from three different water flows and classified in three groups: sediments from Carrion River (group I), sediments from the northern stream (group II) and sediments from the southern stream (group III). Group I included SD-1, SD-2 and SD-3 samples (Figure 1), and were selected in relation with the CFPP and active mine locations: SD-1 was collected before the area with coal cycle activities, SD-2 was collected immediately past CFPP and SD-3 was chosen far away from the working area. These river sediment samples were collected taking wet solid material from the river bed in a way such that perturbations of river flux should be considered not important. Group II included SD-4 to SD-7 samples (Figure 1). They were collected from a stream located near the most important mine in the area. Soil reference sample S-1 was taken in the vicinity of group II samples, particularly close to SD-5. Group III includes SD-8 and SD-9 samples (Figure 1). They were collected from an area without mining activity and the reference soil sample S-2 was taken close to SD-8. Stream sediment samples were usually collected from deep deposited material in non-running water points, because these streams present high seasonal variations with high flows in periods of a few weeks, and zero flow during the greater part of the year. Both sampled streams flow into Carrión River. Sixty-seven surface soil samples (5 cm depth) were collected around the CFPP in a square of 8 Km x 8 Km. The sampling sites were chosen from non-cultivated public areas and undisturbed flat grasslands, free of trees or other obstacles. The distance and height of the sampling points with respect to the CFPP, ranged from 782 m to 4806 m and from 5 m to 436 m, respectively. We have selected thirteen sampling points when "cores" of soil were collected. Cores were taken using a stainless steel cylinder sampler of 4.5 cm diameter and 50 cm length. The cylindrical vertical soil sample was sliced in situ at 5 cm intervals. Finally, all samples were air-dried when necessary and oven dried at 110 ºC overnight. They were sieved, at a 2 mm grain size, homogenised and packaged in Petri dishes and stored at least during 3 weeks, in order to preclude Radon emanation and also to ensure secular equilibrium of ^{226}Ra and ^{214}Pb.

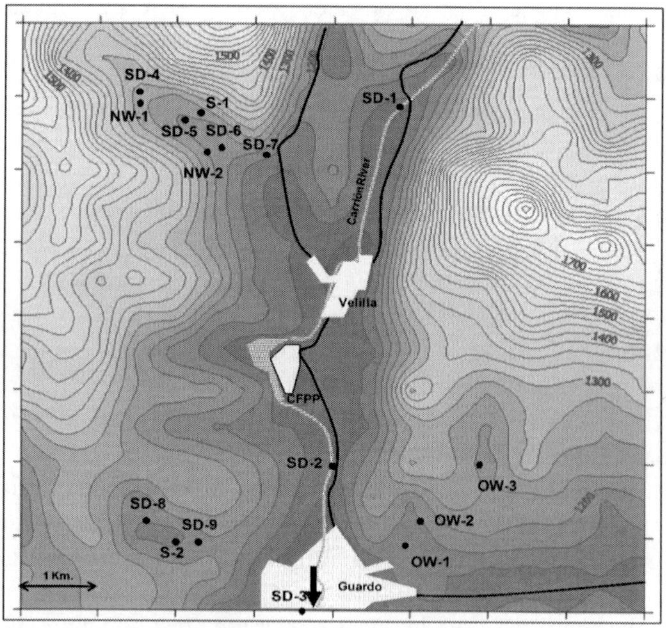

Figure 1. Map of sampling points in the studied area: wastes (W), sediments (SD) and soils (S).

The activity concentrations of ^{238}U, ^{226}Ra, ^{224}Ra, ^{210}Pb, ^{232}Th, ^{40}K and ^{137}Cs were determined by direct gamma-ray spectrometry using a Canberra n-type HPGe detector, with 25% relative efficiency, resolution of 1.1 keV at 122 KeV and of 2.0 KeV at 1.33 MeV, where self-absorption correction was applied [13]. Uncertainty (with k=1) was determined attending to statistical uncertainty of peak areas provided by Canberra Genie-2000 software, uncertainty values calculated from internal procedures for mass and efficiency and literature data for gamma emissions parameters, always according to ISO accepted rules. Limits of detection were estimated less than 1 Bq Kg^{-1} for all radionuclides.

Statistical analysis has been performed with STATGRAPHICS CENTURION XVI software. The Kendall τ rank test was carried out with the Wessa Free Statistics Software, Ordinary kriging, a geostatistical technique to interpolate the value of a parameter at an unobserved location from observations of its values at nearby locations, has been used to build maps of the radionuclide activities and other variables, in relation with the positions with respect to CFPP. For this purpose, GOLDEN SOFTWARE SURFER 9 © software has been used.

3. Results and Discussion

3.1. Thorium in Coal and Ash

The average coal activity concentrations for ^{238}U, ^{226}Ra, ^{232}Th and ^{40}K were 24, 30, 23 and 242 Bq Kg^{-1}, respectively (Table 1). These values were found close to the world average values for coal, according to UNSCEAR [1] and to those found in the literature (Table 1). The highest concentrations of Uranium and Radium are found in lignite from mines in Greece [4, 8, 10], Brazil [3] and Turkey [15], being Thorium concentration in lignite and coal very closed. Our data show that the coal used in Velilla CFPP is not lignite.

The analysis of the radionuclide activity concentration in ashes is important to estimate (i) the efficiency of the burning process and (ii) the potential risk of the use of these ashes as part of new material, with a special attention to the radon emanation. Table 1 reports the measured activity concentration for Thorium and the other radionuclides in ash that were always higher than the ones found in coal, and been our values similar to those reported by some authors in Table 1. This fact can be explained because the ash is the result of the combustion process of the coal, which involves an enrichment in heavy metals and radionuclides in the ash.

Table 1. Comparative activity concentrations (in Bq Kg^{-1}) for coal and ash samples

Reference	Coal				Ash				
	^{238}U	^{226}Ra	^{232}Th	^{40}K	Type	^{238}U	^{226}Ra	^{232}Th	^{40}K
This work	20 -30	22 - 37	15 - 30	148 - 310	Fly	103	128	88	860
Aytekin & Baldik (2011) [14]	-	29-61	32-55	229-414	Fly	-	80-98	64-85	754-992
Parmaksiz et al. (2011) [15]	-	199.8	32.0	152.8	Fly	-	431.5	87.3	439.0
					Bottom	-	380.3	74.0	401.3
Papastefanou (2010) [4]	117-399	44-236	9-41	59-227	Fly	263-950	142-605	27-68	204-382
Psichoudaki and Papaefthymiou(2008) [6]	84	93	19	79	Fly	294	276	51	270
Mahur et al. (2008) [16]	-	-	-	-	Fly	85 - 126	-	98 - 141	267-365
Flues et al. (2007) [3]	18 - 48	27 - 35	20 - 30	181 - 433	Fly	64	62	52	534
Flues et al. (2007) [3]	159 - 807	135 - 698	15 - 32	136 - 245	Fly	1424	1284	88	764
Cevik et al. (2007) [17]	-	14	11	123	Fly	-	149	58	94
Karangelos et al. (2004) [10]	306	346	19	173	Fly	964	904	53	454
					Bottom	681	662	44	405
Mishra (2004) [18]	-	11 - 66	19 - 93	15 - 444	Fly	-	41 - 152	96 - 178	148-840
Bem et al. (2002) [19]	23.5	18.1	14.3	129.9	Fly	134.2	94.1	74.7	646.9
					Bottom	65.8	62.2	52.6	467.0
UNSCEAR (2000) [1]	10-25	-	10-25	-	Fly	200-400	-	200	-
Alvarez & Dopico (1998) [20]	3 - 67	-	0.1 – 6.5	26 - 2490	-	-	-	-	-
Jamil et al. (1998) [21]	-	7.7 – 31.4	-	17.1– 21.4	-	-	-	-	-
Tso and Leung (1996) [22]	-	17	20	24	Fly	-	140	155	178
					Bottom	-	100	105	132
Fardy et al. (1989) [23]	25	21	24	75	Fly	89	79	91	364

Reference	Coal				Ash				
	^{238}U	^{226}Ra	^{232}Th	^{40}K	Type	^{238}U	^{226}Ra	^{232}Th	^{40}K
Simopoulos and Angelopoulos (1987) [8]	-	321	21	191	Fly	-	845	56	502
					Bottom	-	587	44	423
Tadmor (1986) [24]	10 - 160	10 - 260	2 -100	-	Fly	20 - 700	20 - 500	20 - 100	-
Tracy and Prantl (1985) [25]	12.4	-	7.5	26.4	Fly	92	-	58	204
Nakaoka et al. (1984) [26]	12.4	-	13.3	72.2	Fly	117.5	-	126.8	687
Mishra et al. (1980) [18]	-	15 - 48	29 - 44	-	Fly	-	52- 152	78 - 178	-
Tomczynska et al. (1980) [27]	37.9	-	30.0	294	Fly	96.7	-	73.9	728
Coles et al. (1978) [28]	8.9	7.4	6.3	27.0	Fly	70.3	85.1	62.9	299.7
					Bottom	55.5	70.3	55.5	251.6

Table 2. Enrichment factors (E.F.) for several radionuclides

Reference	Type	^{238}U	^{226}Ra	^{210}Pb	^{232}Th
This work	Fly	1.2	1.2	0.9	1.1
Aytekin and Baldik (2011) [14]	Fly	-	2.2	-	1.7
Parmaksiz et al. (2011) [15]	Fly	-	0.8	-	0.8
	Bottom	-	0.7	-	0.9
Psichoudaki and Papaefthymiou (2008) [6]	Fly	1.0	0.9	-	0.8
Karangelos et al. (2004) [10]	Fly	1.2	1.1	1.2	1.1
	Bottom	0.9	0.8	0.3	1.0
Bem et al. (2002) [19]	Fly	1.2	1.0	1.3	1.1
	Bottom	0.8	1.0	0.4	1.0
Tso and Leung (1996) [22]	Fly	-	1.1	-	1.0
	Bottom	-	1.0	-	1.0
Fardy et al. (1989) [23]	Fly	0.7	0.8	0.7	0.8
Simopoulos and Angelopoulos (1987) [8]	Fly	-	1.0	-	1.0
	Bottom	-	0.8	-	0.9
Tracy and Prantl (1985) [25]	Fly	1.0	-	-	1.0
Nakaoka et al. (1984) [26]	Fly	1.0	-	-	1.0
Tomczynska et al. (1980) [27]	Fly	1.0	-	-	1.0
Coles et al. (1978) [28]	Fly	0.7	1.0	0.5	0.9
	Bottom	0.7	1.0	0.2	0.9

All the values have been calculated for this work using data reported by the authors.

An interesting parameter can be evaluated, the enrichment factor (E.F.) defined [4, 15, 16, 18] as:

$$E.F. = \frac{[X]_{ash} / [^{40}K]_{ash}}{[X]_{coal} / [^{40}K]_{coal}} \tag{1}$$

where [X] is the activity concentration of a particular radionuclide. The parameter is evaluated taking ^{40}K for normalization, because that isotope can be considered homogeneous in all types of ashes [28]. The data obtained in this work are reported in Table 2 along the E.F. values calculated by us using the average values for coal and ashes published by other authors. E.F. values greater than unity denote enrichment, while values lower than unity show depletion. Our fly ash vales showed enrichment in ^{238}U, ^{226}Ra, and ^{232}Th, and are close to other reported values [8, 10, 19, 22]. Note that the parameters for ^{232}Th and ^{226}Ra reported by Aytekin and Baldik [14] are particularly high when compared with the rest of the data in the table. No enrichment was found for ^{210}Pb. The enrichment and volatilization behaviour of trace elements during coal combustion is determined by the physicochemical properties of elements, the nature of the burning process and the mechanisms occurring in the control devices of the emissions. The enrichment factor in fly and bottom ash is related with the particle size. ^{238}U and ^{226}Ra are heavier than Pb, so Pb goes out of the oven and appears rarely in bottom ash, where its E.F. is usually lower than unity (see other data reported in Table 2). On the contrary, E.F. values greater than unity show the capture achieved by the modern filters used in the CFPP's. Therefore, the analysis of the fly ash data in our study is indicative of an incorrect operation of the filters and precipitators of Velilla CFPP that can originate the releasing to environment of significant amounts of ^{210}Pb activity, which can be easily inhalable and transportable in the atmosphere. This atmospheric fallout of ^{210}Pb on the top layer of soil can contribute to the contamination of vegetation [29]. However, Thorium enrichment factor is close to 1 in our sample and in the rest of the data reported by other authors, that means that no contamination by emission to the atmosphere is expected.

Other parameter useful for the analysis of the radiological impact is the radium equivalent activity (Ra_{eq}). It is a common radiological index [30] used to compare the specific activities of materials containing different amounts of ^{226}Ra, ^{232}Th and ^{40}K and therefore, to evaluate the radiation hazards associated with these radionuclides. The Thorium concentration has an important contribution to evaluate this parameter. Ra_{eq} is related to both the external and internal gamma dose due to radon and its daughters. Several authors [31, 32] have estimated that 370 Bq/Kg of ^{226}Ra, 260 Bq/Kg of ^{232}Th and 4810 Bq/Kg of ^{40}K produce the same gamma-ray dose rate. The index is calculated by using the following expression:

$$Ra_{eq} = [^{226}Ra] + 1.43 \cdot [^{232}Th] + 0.077 \cdot [^{40}K] \quad (2)$$

The maximum allowed Ra_{eq} value in building materials and products for safe use, should be less than 370 Bq/Kg in order to keep the external dose below 1 mSv/y [33]. The value of this parameter for the fly ash collected in Velilla CFPP to be used in building materials is 320 Bq/Kg, only a 13% lower than the maximum allowed value, where the 39% of the total activity is the contribution of thorium.

3.2. Thorium Mobility from the Waste Rock Piles

The activity concentration of the different radionuclides in two sets of coal washing wastes piles, NW (new wastes) and OW (old wastes), are reported in Table 3. The two groups showed clear differences in their Uranium, Radium and Lead contents, being the Thorium and Potassium values very similar in both groups of samples. ^{238}U, ^{226}Ra, ^{224}Ra and ^{210}Pb activity values in old wastes were lower than the found in new ones, thus meaning that these three elements have more mobility than ^{232}Th and ^{40}K. This difference was also found when using the ratio between the averages activity concentrations in new and old wastes (NEW/OLD) for every radionuclide. These results are collected in Table 3, which shows that ^{232}Th and ^{40}K activity concentrations have no variations along time, whereas the ^{238}U, ^{226}Ra and ^{210}Pb activity values have decreased with the years. The mobility of these isotopes can be explained on the basis on their different solubility in water.

The ratios of the different radionuclides are also reported in Table 3. The value of [^{238}U/^{232}Th] ratio was very different in the two groups of samples, with the values for new wastes being twice than the corresponding to old ones, and also that the values for NW-1 (1.28) and NW-2 (0.97) were very close to those found in coal (1.03). On the other side, the values for ratios [^{238}U/^{226}Ra] and [^{210}Pb/^{226}Ra] showed no clear differences between the two groups of samples, and were very close to those found in coal, 0.80 and 1.37, respectively. Then, the results seem to indicate that, in the wastes, ^{238}U has been preferentially leached from the parent material with the rainfall as the most important leaching agent. The migration of uranium could affect preferentially to nearest streams, whose are sediments could generate a local increment of radioactivity. However, no migration of Thorium is expected, and thus, this radiological impact can be only expected to be in the waste materials.

The rock of mine (ROM) coal radioactivity contents (in Bq Kg^{-1}) can be estimated, on the basis of the available data, using the average activity concentrations for each radionuclide ([X]) in washed coal and in new wastes, and assuming that a 20% of mass was lost during the washing process:

$$[X]_{ROM} = 0.2 \cdot [X]_{waste} + 0.8 \cdot [X]_{coal} \tag{3}$$

The estimated activity concentration in ROM coal are reported in Table 3, which shows that the corresponding ratio values are close to those found in coal. Taking into account the mass loss, the radionuclide contents are greater in the washed material than in the original coal. The Thorium activity in the ROM is 30, and in the wastes gets to the double.

The analysis of the ratios between the activity concentration for two particular isotopes let us know if the secular equilibrium is happening, which can be observed when values different from unity are found. Usually, all decay products of ^{238}U and ^{232}Th in the coal rock of mine (ROM) are in radioactive equilibrium with their precursors [34]. The average values of [^{238}U/^{226}Ra] and [^{210}Pb/^{226}Ra] ratios in ROM were 0.80 and 1.37, respectively. On the other hand, the data have also shown the same equilibrium for the uranium ore elements in the coal samples provided by CFPP (the ratio between Uranium and Radium is also 0.80) but different concentration of the different isotopes. The average activity concentration of ^{238}U and ^{226}Ra in the coal samples (Table 1) are lower than the ROM coal ones (Table 3) and this feature can be originated by a previous coal wash. Given the low contents found in these coal samples, a significant concentration of some radionuclides can be expected in the waste rock piles and water.

Given the potential hazard the waste materials could pose for the environment, we have calculated the air absorbed dose rate (D) originated by the mean activity concentrations of the studied natural radionuclides in the wastes. According to UNSCEAR [1], dose rate D (in nGy h^{-1}) could be calculated using the activity values in Bq Kg^{-1} for selected radionuclides, as:

$$D = 0.462 \cdot [^{238}U] + 0.604 \cdot [^{232}Th] + 0.0417 \cdot [^{40}K] \tag{4}$$

The D values are collected in Table 3, and were always higher than 60 nGy h^{-1}, the world average dose rate [1]. This fact implies that waste piles, despite being very localized points, can contribute to increase the dose rate in the surrounding area. Given that Thorium remains in the waste piles, their contribution has to be taken into account.

The activity of Thorium in the wasted piles ranges from 50 to 66, where no difference have been found between new and old waste piles, and being their contribution to the D values of 36-39%.

Table 3. Activity concentrations (in Bq Kg^{-1}), activity ratios and air absorbed dose rate D (nGy h^{-1}) for wastes samples and ROM coal (estimated values)

Sample	^{238}U	^{226}Ra	^{224}Ra	^{210}Pb	^{232}Th	^{40}K	^{238}U / ^{226}Ra	^{210}Pb/ ^{226}Ra	^{238}U / ^{232}Th	D
NW-1	64 ± 5	82 ± 6	60 ± 4	89 ± 7	50 ± 4	590 ± 40	0.78	1.09	1.28	84
NW-2	64 ± 5	80 ± 6	78 ± 6	96 ± 7	66 ± 5	800 ± 60	0.80	1.20	0.97	103
NEW [(a)]	64	81	69	92	58	695	0.79	1.14	1.10	94
OW-1	28 ± 2	42 ± 3	58 ± 4	49 ± 4	50 ± 4	660 ± 60	0.67	1.17	0.56	71
OW-2	30 ± 2	39 ± 3	64 ± 5	44 ± 4	53 ± 4	750 ± 50	0.77	1.13	0.57	77
OW-3	33 ± 3	41 ± 3	71 ± 5	72 ± 6	61 ± 4	860 ± 60	0.80	1.76	0.54	84
OLD [(b)]	30	41	64	55	55	757	0.73	1.34	0.54	77
NEW / OLD [(c)]	2.1	2.0	1.1	1.7	1.1	0.9	-	-	-	-
ROM Coal[(d)]	32	40	36	51	30	330	0.80	1.28	1.07	47

[(a)] Average for NW-1 and NW-2.

[(b)] Average for OW-1, OW-2, OW-3.

[(c)] Ratio from average values.

[(d)] Estimated rock of mine coal.

Table 4. Activity concentrations (in Bq Kg^{-1}) and activity ratios for bulk sediment samples (SD-1 to SD-9) and soil samples (S-1 and S-2)

Samples	SD-1	SD-2	SD-3	SD-4	SD-5	SD-6	SD-7	SD-8	SD-9	S-1	S-2
^{238}U	17	18	14	27	29	29	28	28	29	32	34
^{226}Ra	23	23	17	36	41	38	38	33	37	48	38
^{210}Pb	41	34	24	42	46	40	47	40	43	139	101
^{232}Th	26	24	20	11	22	25	26	32	45	40	51
^{40}K	321	270	211	107	231	276	307	406	527	410	660
$[^{238}U / ^{226}Ra]$	0.73	0.78	0.80	0.74	0.70	0.77	0.74	0.83	0.78	0.67	0.89
$[^{210}Pb / ^{226}Ra]$	1.76	1.44	1.35	1.15	1.12	1.06	1.23	1.20	1.16	2.90	2.66
$[^{238}U / ^{232}Th]$	0.64	0.76	0.69	2.52	1.35	1.15	1.08	0.85	0.64	0.80	0.67

Table 5. Activity concentration (in Bq Kg^{-1}) for radionuclides in every 5 cm of depth

Depth		0 – 5 cm.	5 – 10 cm.	10 – 15 cm.	15 – 20 cm.	20 – 25 cm.	25 – 30 cm.
^{238}U	a)	20 – 43	20 – 43	19 – 41	21 – 50	20 – 45	26 – 43
	b)	34 (7)	35 (7)	34 (7)	37 (10)	35 (7)	35 (5)
^{226}Ra	a)	29 – 63	28 – 70	29 – 72	27 – 71	24 – 71	29 – 79
	b)	44 (10)	46 (11)	46 (11)	47 (12)	45 (12)	48 (14)
^{210}Pb	a)	45 – 177	41 – 84	38 – 65	25 – 55	22 – 60	32 – 60
	b)	102 (30)	59 (12)	50 (10)	43 (8)	41 (10)	42 (8)
^{232}Th	a)	31 – 64	34 – 73	33 – 70	30 – 75	30 – 74	38 – 76
	b)	50 (10)	52 (12)	52 (11)	53 (13)	54 (13)	55 (11)
^{224}Ra	a)	40 – 77	41 – 84	42 – 87	35 – 89	35 – 89	45 – 88
	b)	60 (12)	64 (15)	64 (14)	63 (15)	64 (16)	63 (13)
^{137}Cs	a)	0 – 85	0 – 31	0 – 20	0 – 15	0 – 10	0- 4.6
	b)	28 (21)	18 (9)	12 (6)	6 (5)	2 (3)	1.0 (1.5)
^{40}K	a)	250 – 790	260 – 930	230 - 920	230 – 940	200 – 860	260 – 290
	b)	507 (163)	546 (199)	531 (191)	522 (188)	512 (188)	549 (181)

a) Range. b) Average and standard deviation.

3.3. Migration of Thorium to Sediments

The importance of the study of the sediments is related with the availability, i.e. mobility and solubility, of the natural radionuclides in water. The study of the contamination by natural radionuclides in Spanish river sediments and waters has been performed by several authors [35-37]. However, no data have been reported for the presence of radionuclides in sediments caused by releasing from CFPP waste piles.

The activity concentrations of ^{238}U, ^{226}Ra, ^{210}Pb, ^{232}Th and ^{40}K in nine sediment (SD-1 to SD-9) and two chosen soil (S-1 and S-2) samples, along the corresponding [$^{238}U/^{226}Ra$], [$^{210}Pb/^{226}Ra$] and [$^{238}U/^{232}Th$] ratios are reported in Table 4. As can be seen in that Table, activity concentrations for all the radionuclides are very similar in the eleven samples, exception made from the slightly inferior data reported for Uranium in the SD-4 sample and the data for Lead, in both soil samples. The ratio [$^{238}U/^{232}Th$] are always lower than 1, with the exception of the group II (SD-4 to SD-7), being the main discrepancy in the case of SD-4 sample. The values found for the ratio [$^{210}Pb/^{226}Ra$] in the eleven samples were quite different for soil samples, S-1 and S-2. The differences observed for ^{210}Pb activity concentration and [$^{210}Pb/^{226}Ra$] ratio between the sediments and soils data, can be explained in terms of the competition of different processes, such as mobility, solubility and deposition air to soil. The ^{210}Pb is a product of the disintegration of the gaseous ^{222}Rn, and it is incorporated to the soil by an air-soil deposition mechanism. This behaviour is difficult to prove when the samples are perturbed, as occurs in the case of sediments. In this case, the variability in the activity concentration of some elements is better explained by solubility and mobility processes. The contents in ^{232}Th and the [$^{238}U/^{232}Th$] ratio are quite different in SD-4, a fact that could be explained by a migration of uranium from the waste piles located in the vicinity of the sediment SD-4. A better analysis can be done using the ratios [$^{238}U/^{226}Ra$], [$^{210}Pb/^{226}Ra$] and [$^{238}U/^{232}Th$] shown in Table 4. These ratios behaved in the same way than the previously discussed bulk samples. The [$^{238}U/^{226}Ra$] and [$^{210}Pb/^{226}Ra$] ratios were found close to 1, with some small variations, for all the sediment samples. The same can be said for the [$^{238}U/^{232}Th$] ratio, with the obvious exception of SD-4, whose value, higher than 2, indicates a strong disequilibrium due to a large uranium concentration probably released from the nearby waste piles.

3.5. Vertical Distribution of Thorium in Soil

The general study of the thirteen profiles (activity as a function of depth) for every radionuclide has been done, where the average activity if every radionuclide for each fraction of 5 cm is used, finally a total of 72 samples have been measured. The statistical values (minimum, maximum, average and standard deviation) of the ^{238}U, ^{226}Ra, ^{210}Pb, ^{232}Th, ^{224}Ra, ^{40}K and ^{137}Cs activities are reported in Table 5. The range of activities in Bq $Kg^{-1}$of the data is 20 - 50 for ^{238}U, 24 - 79 for ^{226}Ra, 22 - 177 for ^{210}Pb, 30 - 76 ^{232}Th, 35 – 89 for ^{224}Ra, 200 - 940 for ^{40}K, and 0- 85 for ^{137}Cs. These values are accepted as normal for soils for natural radionuclides corresponding to 35, 35, 30 and 400 Bq Kg^{-1} for ^{238}U, ^{226}Ra, ^{232}Th and ^{40}K, respectively [1]. Upper values than ours are those for ^{238}U and ^{232}Ra reported by Flues et al [38] and Papp et al [5], which are related with the use of uraniferous coal in the CFPP. Similar values to ours are those reported by some authors [19, 2, 39]. For ^{137}Cs and ^{210}Pb activities, the highest variability is observed in the first phase (0-5 cm) (see Table 5). This fact can be due to several factors as topography, soil type, etc. and it is also observed in other studies.

All the radionuclides, with the exception of Caesium and Lead, show a constant distribution along the depth. The average concentration of ^{238}U is lower than the ^{226}Ra one in all the phases, and this feature suggests that there is not a radioactive equilibrium. This disequilibrium cannot be attributed to Radium deposition from airborne discharges because in a such case, a higher concentration in the upper layer with reference to the deeper ones should have been detected. A similar distribution along depth can be observed for ^{232}Th and ^{40}K activities. This behavior, with the isotopes having not connection by disintegration, can be attributed to the very similar physicochemical properties of the corresponding ions.

A very similar profile was also found showed by ^{210}Pb and ^{137}Cs, whose radioactivity concentration decrease with depth, because both radionuclides appear in the top of soil by deposition from air, despite of their different origin (natural or artificial respectively for Pb and Cs). As could be expected, ^{137}Cs activity values become zero at a 2530 cm depth, whereas ^{210}Pb reaches a near constant value, very close to the activity corresponding to its father ^{238}U.

Table 6. Ratios of activity values for the seventy-two measured samples

Ratio	Depth(cm)	C-1	C-2	C-3	C-4	C-5	C-6	C-7	C-8	C-9	C-10	C-11	C-12	C-13
	0 – 5	0.95	1.41	1.13	1.34	1.33	1.29	1.29	1.11	1.02	1.37	1.37	1.54	1.75
	5 – 10	1.33	1.05	1.22	1.28	1.27	1.16	1.30	1.07	1.19	1.57	1.57	1.70	1.60
	10 – 15	1.22	1.16	1.15	1.08	1.53	1.26	1.24	1.15	1.44	1.61	1.61	1.64	1.79
[$^{226}Ra/^{238}U$]	15 – 20	-	1.10	1.22	1.57	1.17	1.31	1.13	0.90	1.26	1.30	1.30	1.42	1.67
	20 – 25	-	1.24	1.21	1.29	1.20	1.33	1.13	1.00	1.14	1.50	1.50	1.53	1.43
	25 – 30	-	1.31	-	-	-	1.40	1.10	1.25	1.41	1.35	1.35	1.43	1.35
	0 – 5	2.88	2.20	2.63	1.81	2.94	2.58	2.39	4.32	1.89	1.71	1.71	1.63	1.29
	5 – 10	1.58	1.36	0.91	1.32	1.68	1.46	1.12	1.13	1.32	1.22	1.22	1.13	1.47
	10 – 15	0.98	1.39	0.91	1.05	1.41	1.41	0.88	0.96	0.93	1.07	1.07	1.05	1.12
[$^{210}Pb/^{226}Ra$]	15 – 20	-	0.98	0.82	0.85	0.93	1.05	0.84	0.93	0.93	0.91	0.91	0.94	0.91
	20 – 25	-	0.89	0.79	0.89	0.92	1.13	0.87	0.91	0.95	0.89	0.89	0.92	0.73
	25 – 30	-	0.83	-	-	-	1.05	1.00	0.94	0.91	0.71	0.71	0.79	0.91
	0 – 5	0.78	0.62	0.68	0.78	0.77	0.58	0.66	0.59	0.93	0.67	0.67	0.61	0.51
	5 – 10	0.63	0.84	0.55	0.70	0.65	0.65	0.60	0.59	0.82	0.61	0.61	0.60	0.53
	10 – 15	0.51	0.75	0.68	0.72	0.58	0.61	0.64	0.71	0.64	0.63	0.63	0.63	0.51
[$^{238}U/^{232}Th$]	15 – 20	-	0.74	0.79	0.53	0.77	0.62	0.68	0.67	0.78	0.67	0.67	0.69	0.54
	20 – 25	-	0.66	0.61	0.61	0.67	0.57	0.65	0.61	0.72	0.58	0.58	0.62	0.64
	25 – 30	-	0.58	-	-	-	0.56	0.68	0.53	0.62	0.66	0.66	0.67	0.68

In order to study the contamination of the area, the ratios of activity [^{226}Ra/^{238}U], [^{210}Pb/^{226}Ra] and [^{238}U/^{232}Th] have been evaluated, since these ratios are a good index to examine the disequilibrium between radionuclides belonging to a same radioactive series. The values for all the samples were calculated and are reported in Table 6. The average activity values [^{226}Ra/^{238}U], [^{210}Pb/^{226}Ra] and [^{238}U/^{232}Th] were 1.27, 1.31 and 0.66, respectively. The deviation of the values from the unity, is an evidence of a disequilibrium. The first ratio, [^{226}Ra/^{238}U], that takes into account the Uranium elements series, presents most of the values higher than 1. Although a variability data (ratio values vary between 0.83 and 1.79) can be observed, the average seems to show a constant distribution with depth, so the disequilibrium must be explained attending to soil types or local variations. In the case of [^{210}Pb/^{226}Ra] ratio, average values for in every selected fraction of 5cm. of depth were 2.42, 1.35, 1.14, 0.95, 0.90 and 0.92 consecutively. These values are related with a ^{210}Pb excess respect from the basis of their fathers in the Uranium series, and this descendant trend is observed until 20 cm depth, where a constant value of the concentration of this isotope is found, very close to the ^{226}Ra activity. This behaviour has been also observed by other authors [38], which found values for these ratios between 1.25 and 1.89. Finally, the [^{238}U/^{232}Ra] ratio values vary between 0.51 and 0.93 and clearly show the normal concentration of natural radionuclides in soils, with no evidence of contamination of the soils by the CFPP releases.

Deposition effects for natural radionuclides have been studied by some authors [19, 2, 38] using the *K*-parameter. This parameter is estimated for each radionuclide, taking into account the activity of the radionuclide [X] in the horizons A and B (ranging respectively from 0-15 cm and 15-30 cm in the core sample), as follows:

$$K_X = [X]_{HorizA} / [X]_{HorizB} \quad (5)$$

Values of *K* different than unity mean than a profile for a radionuclide can be found. This parameter cannot be used with ^{137}Cs because this element does not exist in the deepest fractions, given its artificial nature and its atmospheric origin. The estimated K values for our sampled points are showed in Table 7 for selected radionuclides. With the exception of ^{210}Pb, all the *K* values range between 0.8 and 1.2, with average values close to 1. However, K_{Pb} values range between 1.5 and 2.6, indicating a decreasing trend with depth. The K-parameter for Thorium for each profile is the unit, that means that no migration of thorium is observed, which is related with the low solubilty in water of this element, as we pointed out before in this work.

3.5. Thorium Content in Surface Soil Samples

The activities of six natural radionuclides (^{238}U, ^{226}Ra, ^{210}Pb, ^{232}Th, ^{224}Ra , ^{40}K) and the artificial one ^{137}Cs have been measured in sixty-seven top soils. The ^{137}Cs activity was measured because this radionuclide can be used as a good tracer to study soil deposition processes [5]. Table 8 shows the mean and the rank activity values found for the radionuclides measured in the present work. The activity, in Bq kg^{-1}, varies from 11 to 50 for ^{238}U, 14 to 67 for ^{226}Ra, 15 to 68 for ^{232}Th, 17 to 29 for ^{210}Pb, 380 to 78 for ^{224}Ra and 97 to 790 for ^{40}K. For the artificial ^{137}Cs radionuclide, the values range from very low - below the detection limit – values up to 209 Bq Kg^{-1} .With the exception of ^{210}Pb and ^{137}Cs, the data distributions are normal according to the Anderson-Darling normality test (p-value greater than 0.05).

Table 7. K-parameter value for radionuclides for all sampling points

K	^{238}U	^{226}Ra	^{210}Pb	^{232}Th	^{224}Ra	^{40}K
C-1	1.1	1.0	1.9	0.9	1.0	1.0
C-2	1.0	1.0	1.8	0.9	0.9	0.9
C-3	0.8	0.8	1.5	0.9	0.9	1.0
C-4	1.0	0.9	1.5	0.8	0.9	0.9
C-5	1.0	1.2	2.6	1.1	1.2	1.2
C-6	1.0	0.9	1.5	0.9	0.9	1.0
C-7	1.0	1.1	1.7	1.0	1.1	1.1
C-8	0.9	1.0	2.1	0.9	0.9	1.0
C-9	1.1	1.0	1.5	1.0	1.0	1.0
C-10	0.9	1.0	1.9	1.0	0.9	1.0
C-11	1.0	1.0	1.7	1.0	1.0	0.9
C-12	1.0	1.1	1.5	1.0	1.0	1.1
C-13	0.8	1.0	1.5	1.0	1.0	1.0
Average	1.0	1.0	1.7	1.0	1.0	1.0
St. Dev.	0.1	0.1	0.3	0.1	0.1	0.1

Table 8. Average and rank values (in Bq kg^{-1}) of the activities of radionuclides measured in soils around CFPP found in this work and in the literature

	^{238}U		^{226}Ra		^{232}Th		^{40}K	
	mean	rank	mean	rank	mean	rank	mean	rank
This work	30	11-50	39	14-67	43	15-68	445	97-790
UNSCEAR 2000[1]	35	-	35	-	30	-	400	-
Gür and Yaprak 2010[40]	-	-	32	18-53	37	17-89	455	203-794
Gür and Yaprak 2010[40]	-	-	42	9-168	32	6-74	365	117-937
Nenadovic et al 2012[41]	-	28-44	-	-	-	59-71	-	335-517
Psichoudaki and Papaefthymiou 2008[6]	42	25-55	27	20-30	36	26-45	496	404-571
Dai et al. 2007[2]	-	-	32	24-40	50	38-66	720	498-858
Dowdall et al. 2004[39]	32	17-49	36	15-62	24	9-43	362	115-614
Bem et al. 2002[19]	-	12-22	-	10-21	-	9-20	-	224-427
Flues et al. 2002[38]	-	24-78	-	18-84	-	14-51	-	71-328
Papp et al. 2002[5]	41	0-1346	43	21-1256	24	15-42	330	176-567
Rouni et al. 2001[7]	-	-	-	23-337	-	12-43	-	154-631
Bem et al. 1998[42]	-	7-26	-	14-38	-	15-32	-	274-503
Papastefanou et al. 1988[43]	48	17-360	200	50-941	-	-	-	-
Bunzl et al. 1984[44]	-	-	-	13-58	-	22-48	-	259-703
Rosner et al. 1984[45]	31	15-67	35	11-59	27	11-48	362	244-703

Table 8 also shows comparative data found in the literature. The results reported in our study are very close to those given by UNSCEAR [1] for soils, (35 Bq kg^{-1} for ^{238}U and ^{226}Ra, 30 Bq Kg^{-1} for ^{232}Th and 400 Bq Kg^{-1} for ^{40}K), and also to those measured recently in soils around a CFPP in Serbia [40, 41], and also to the found by other authors [2, 6, 19, 38, 39, 42, 44, 45]. Most of these authors agree that ash releases have not a significant impact over natural background (even if lignite was used) because of the efficiency of the filtering system used nowadays, which reaches a 99.5% of efficiency [19]. However, data from Papp et al. [5] and Papastefanou et al. [43] clearly show higher values.

It is important to mention that Velilla's CFPP does not use uraniferous coals, which sometimes appear related with the existence of hot spots of pollution, particularly for ^{226}Ra, on the soils around a CFPP as it has been observed by some authors [43, 5, 7]. On the other hand, the studies on dispersion models of fly ash, show that high radionuclide levels in the soils are limited to distances lesser than 1 km from CFPP [2, 38, 4].

As discussed before, the activity ratios [^{226}Ra/^{238}U], [^{210}Pb/^{226}Ra] and [^{238}U/^{232}Th] can be used as indicator of the radioactive contamination of soils. As the radionuclides of a given daughter-parent ratio belong to the same radioactive series, ratio values different from the unity will indicate a disequilibrium, and therefore the existence of contamination. For [^{210}Pb/^{226}Ra] whose mean value is 3.2, we can see that all the soils present values higher than one, with the excess of ^{210}Pb being caused by its deposition from the air to the topsoil. The others two ratios, [^{226}Ra/^{238}U] and [^{238}U/^{232}Th], with mean values of 1.33 and 0.71 respectively, are also far from the equilibrium. The concentration of uranium is lower than the other two isotopes, and this feature can be explained on the basis of ^{238}U migration processes of that decrease its activity in the soils. The different physicochemical properties of these elements contribute to a better solubility of uranium in water leading to its greater mobility under the effect of the rainfall.

Table 9. (a) Pearson r and (b) Kendall τ rank correlation coefficients for the measured radionuclides with the significant associations highlighted (p-values in parenthesis)

(a) Pearson r coefficients

	^{238}U	^{226}Ra	^{232}Th	^{224}Ra
^{226}Ra	**0.6878**			
	(0.000)			
^{232}Th	**0.8645**	**0.6608**		
	(0.000)	(0.000)		
^{224}Ra	**0.8695**	**0.6378**	**0.9881**	
	(0.000)	(0.000)	(0.008)	
^{40}K	**0.7812**	**0.4851**	**0.9064**	**0.9100**
	(0.000)	(0.000)	(0.000)	(0.000)

(b) Kendall τ rank correlation coefficients

	^{238}U	^{226}Ra	^{232}Th	^{224}Ra	^{137}Cs	^{40}K
^{210}Pb	0.043	0.0630	-0.0261	-0.0224	**0.459**	0.0233
	(0.618)	(0.461)	(0.762)	(0.7950)	(0.000)	(0.787)
^{137}Cs	-0.044	-0.016	-0.135	-0.121	---------	-0.166
	(0.606)	(0.853)	(0.111)	(0.153)	---------	(0.051)

Table 9 shows (a) the Pearson, r, and (b) the Kendall τ rank correlation coefficients of, respectively, the normal and non-normal measured radionuclides, with the significant correlations highlighted. The significant correlations between ^{226}Ra and ^{238}U (r=0.6878; p=0.000) and between ^{238}U and ^{232}Th (r=0.8645, p=0.000) confirm the tendencies seen in the previous paragraph, whereas the non-significant correlation between ^{226}Ra and ^{210}Pb (τ=0.0630; p=0.461) shows again that ^{226}Ra is not related with atmospheric deposition. In addition, to these correlations, that are connected with the daughter-parent ratios in the natural radioactive decay processes. Table 9 also shows other interesting significant positive correlations such the one between ^{210}Pb and ^{137}Cs (τ= 0.459; p=0.000) whose existence, in combination with their common lack of significant correlations with the rest of the measured radionuclides, seem to point towards a common propagation mechanism, surely connected with their atmospheric origin [46]. On the other side, the significant correlations between ^{238}U, ^{226}Ra, ^{232}Th, ^{224}Ra and ^{40}K, also seem to evidence a common origin.

3.6. Risk Assessment

Eisenbud and Petrow [47] were the first authors to point out that the radiation doses, derived from the use of fossil fuels used in power generation, could represent a significant addition to natural radiation doses. The naturally occurring radionuclides present in soil, include ^{238}U and ^{232}Th and all their decay products, as well as ^{40}K and their gamma radiations, represent the main external source of irradiation to the human body [1, 48]. In order to evaluate the hazard to the population living or working near the CFPP, two parameters were evaluated. The first one is the total air absorbed gamma-ray dose rate D (in nGy h^{-1}), and can be calculated with the equation (4). This formula represents the absorbed dose rate in air, at a height of 1m above the ground, emitted from radionuclides present in environmental materials and includes the influence of all decay products of the radioactive series. We have calculated D for the 67 samples, finding results ranging between 18 and 88 nGy h^{-1} with a mean value of 58 ± 18 nGy h^{-1}, very close to 60 nGy h^{-1}, suggested by UNSCEAR [1] as the world usual average value. The distribution of D values in the studied area is shown in Figure 2. The greatest D values (darkest areas) are located in the higher points of the valley, whereas the nearest areas to the CFPP do not show a significant variation of this radioactive natural background. The D small values found [34] and the significant positive correlation between D and the distance to CFPP, again precludes considering this last as the source of radionuclides in the studied area.

The second population risk parameter considered was the annual outdoor effective dose, E, due to soil gamma radiation, calculated as:

$$E = 0.7 \,.\, D \,.\, t \,.\, p \qquad (6)$$

being 0.7 a conversion coefficient, in Sv Gy^{-1}, to transform air doses into effective human body doses. The above formula assumes that people spend outdoors about 20% (outdoor occupancy factor p = 0.2) of their annual exposure time (t =24*365=8760 h). The calculated E value was 71 ± 22 μSv, very similar to the international average value of 70 μSv[1], and its distribution presents a maximum of 108 μSv, similar to others found in the literature and very close to those obtained in Serbia: 90μSv [49] and 96.61μSv [41], although in our case, this maximum value was found in place far away from the CFPP, because the spatial distributions of D and E were found to be identical.

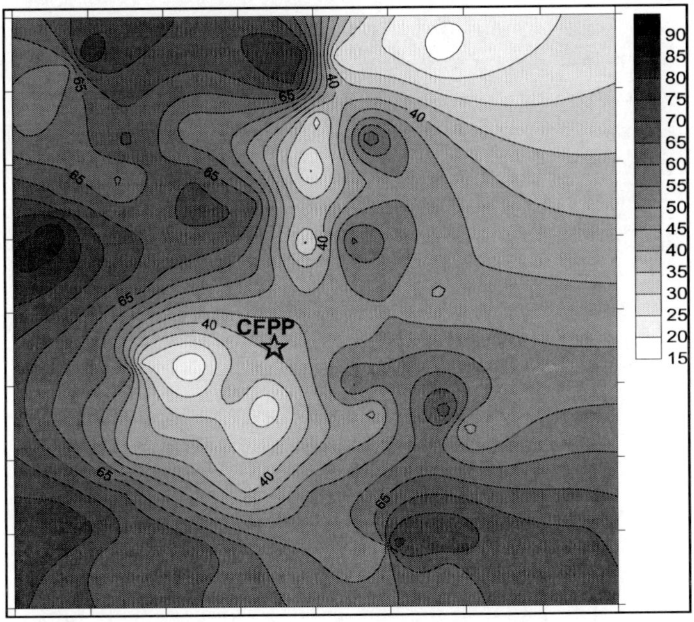

Figure 2. Kriging type estimated map for the distribution of the external dose rate in the studied area (scale of 1 Km).

CONCLUSION

The radioactive contents of ^{232}Th and other radionuclides in different samples have been determined and used to evaluate the distribution and possible impact of a coal fired power plant (CFPP) in Velilla (Spain) on a surrounding area of 64 Km^2. The found values are similar to those shown in literature as background levels, and have been discussed by using, not only the habitual correlations and the daughter-parent ratio analysis, but also powerful and useful statistic tools, and predictive maps, routinely used in environmental studies.

The activity concentrations of ^{232}Th and other radionuclides were measured in several samples in other to determine the radiological impact of the CFPP and the contribution of this isotope. The analyses of the coal used at the Velilla CFPP showed that it has the same concentration in thorium than others. The low concentration of uranium in this coal means that it is a non-uraniferous one and it is not a brown coal or lignite. In the other hand, the analysis of the activity concentration in the fly ash showed an enrichment (attending to E.F. parameter) in ^{232}Th, but not in ^{210}Pb. The mentioned parameter allows us to know the behaviour of radionuclides during coal burning, and indicated a non-correct working of filters and precipitators of the CFPP. In conclusion, a large amount of ^{210}Pb could have been released to environment from stack. Through a comparative analysis of the activity concentration into two groups of washed waste materials (new and old ones), the mobility of ^{232}Th has been found to be lower than that for ^{238}U, ^{226}Ra and ^{210}Pb. The results indicated that, in the wastes, ^{238}U and ^{226}Ra have been preferentially leached from the parent material by the rainfall, given that the low solubility of thorium in water. From this analysis, we can confirm that the impact of the waste piles in the environment is not negligible, given that the value of the dose rate (D) evaluated from every waste pile is always higher than the world average dose rate of 60 nGy h^{-1}. Given that Thorium remains in the piles, its activity has the main contribution to the dose rate emitted. On the other hand, the analysis of the sediment samples lets us confirm that some radioactive elements have migrated from waste piles, but not ^{232}Th. In addition, the activity concentration of thorium in soil samples was measured. The activity concentration of this radionuclide through the profile of the soil confirm the no capacity of Thorium to migrate given its low solubility. Attending to the data found in the surface soil samples, the results show that the radionuclides follows two different models of spatial distribution around the CFPP: a first group, formed by ^{238}U, ^{226}Ra, ^{232}Th, ^{224}Ra and ^{40}K which can be assigned to 'natural background radioactivity', and a second one, composed of ^{210}Pb and ^{137}Cs, and labelled as 'atmospheric fallout radioactivity'. None of these two groups follow a spatial trend suggesting that the CFPP is the source of the measured radionuclides. Furthermore, none of the two indexes estimating the risk to the population, D and E, presents mean values significantly differing from the mean UNSCEAR values, showing the radioactive inocuity of the CFPP at the time of this work. However, all work practices must be aimed at eliminating environmental contamination.

Lastly, this study has also allowed to have a better knowledge of the physicochemical properties and the behaviour in the interface soil-water of thorium, and the roll that it could be have in the environment given its low capacity of migration.

ACKNOWLEDGMENTS

Thanks are due to the "Agencia de Desarrollo Europeo" (ADE), Iberdrola S.A. and "Consejería de Sanidad de Castilla y León" for financial support.

REFERENCES

[1] UNSCEAR. Sources and effects of ionizing radiation. United Nations Scientific Committee on the Effects of Atomic Radiation (UNSCEAR) Report to the General Assembly with annexes, United Nations, New York (2000).

[2] Dai, L., Wei, H., Wang, L. Spatial distribution and risk assessment of radionuclides in soils around a coal-fired power plant: A case study from the city of Baoji, China. *Environ. Res.* 104, 201-208 (2007).

[3] Flues, M., Camargo, I. M. C., Figueiredo, P. M., Silva, P. S. C., Mazzilli, B. P. Evaluation of radionuclides concentration in Brazilian coals. *Fuel* 86, 807-812 (2007).

[4] Papastefanou, C. Escaping radioactivity from a coal-fired power plants (CPPs) due to coal burning and the associated hazards: a review. *J. Environ. Radioact.* 101, 191-200 (2010).

[5] Papp, Z., Dezso, Z., Daroczy, S. Significant radioactive contamination of soils around a coal-fired thermal power plant. *J. Environ. Radioact.* 59, 191-205 (2002).

[6] Psichoudaki, M., Papaefthymiou, H. Natural radioactivity measurements in the city of Ptolemais (Northern Greece). *J. Environ. Radioact.* 99, 1011-1017 (2008).

[7] Rouni, P. K., Petropoulos, N. P., Anagnostakis, E. P., Hinis, E. P., Simopoulos, S. E. Radioenvironmental survey of the Megalopolis lignite field basin. *Sci. Total Environ.* 272, 261-272 (2001).

[8] Simopoulos, S. E., Angelopoulos, M. G. Natural radioactivity releases from lignite power plants in Greece. *J. Environ. Radioact.* 5, 379-389 (1987).

[9] Leopold, K., Michalik, B., Wiegand, J. Availability of radium isotopes and heavy metals from scales and tailings of Polish hard coal mining. *J. Environ. Radioact.* 94, 137-150 (2007).

[10] Karangelos, D. J., Petropoulos, N. P., Anagnostakis, M. J., Hinis, E. P., Simopoulos, S. E. Radiological characteristics and investigation of the radioactive equilibrium in the ashes produced in lignite-fired power plants. *J. Environ. Radioact.* 77, 233-246 (2004).

[11] Ayçik, G. A., Ercan, A. Radioactivity measurements of coals and ashes from coal fired power plants in the southwestern part of Turkey. *J. Environ. Radioact.* 35 (1), 23-25 (1997).

[12] Pandey, V. C., Singh, N. Impact of fly ash in soils system. *Agric. Ecosys. Environ.* 136, 16-27 (2010).

[13] García-Talavera, M., Peña, V. A hybrid method to compute accurate efficiency for volume samples in gamma ray spectrometry. *Appl. Rad. Isot.* 60, 227-232 (2004).

[14] Aytekin H., Baldik R. Radioactivity of coals ans ashes from Catalagzi coal-fired power plant in Turkey. *Radiat. Prot. Dosimetry* doi: 10.1093/rpd/ncr225 (2011).

[15] Parmaksiz A., Arikan P., Vural M., Yeltepe E., Tükenmez I. ^{226}Ra, ^{232}Th and ^{40}K radionuclides enhancement rate and dose assessment for residues of lignite-fired thermal power plants in Turkey. *Radiat. Prot. Dosimetry*, 147 (4), 548-554 (2011).

[16] Mahur, A. K., Kumar, R., Sengupta, D., Prasad, R. Estimation of radon exhalation rate, natural radioactivity and radiation doses in fly ash samples from Durgapur termal power plant, West Bengal, India. *J. Environ. Radioact.* 99, 1289-1293 (2008).

[17] Cevik, U., Damla, N., Nezir, S. Radiological characterization of Cayirhan coal-fired power plant in Turkey. *Fuel* 86, 2509-2513 (2007).

[18] Mishra, U. C., Lalit, B. Y., Ramachandran, T. V. Radioactive release to the environment by thermal power stations using coal as a fuel. *Sci. Total Environ.* 14, 77-83 (1980).

[19] Bem, H., Wieczorkowski, P., Budzanowski, M. Evaluation of technologically enhanced natural radiation near the coal-fired power plants in the Lodz region of Poland. *J. Environ. Radioact.* 61, 191-201 (2002).

[20] Alvarez, M. C., Dopico, M. T. Natural radionuclide contents in Spanish coals of different rank. *Fuel* 77 (13), 1427-1430 (1998).

[21] Jamil, K., Ali, S., Iqbal, M., Qureshi, A. A., Khan, H. A. Measurements of radionuclides in coal samples from two provinces of Pakistan and computation of external gamma rays dose rate in coal mines. *J. Environ. Radioact.* 41 (2), 207-216 (1998).

[22] Tso, M. W., Leung, J. K. C. Radiological impact of coal ash from the power plants in Hong Kong. *J. Environ. Radioact.* 30 (1), 1-14 (1996).

[23] Fardy, J. J., McOrist, G. D., Farrar, Y. J. Neutron activation analysis and radioactivity measurements of Australian coals and fly ashes. *J. Radioanal. Nucl. Chem.* 133, 217-226 (1989).

[24] Tadmor, J. Radioactivity from coal-fired power plants: a review. *J. Environ. Radioact.* 4, 177-204 (1986).

[25] Tracy, B. L., Prantl, F. A. Radiological impact of coal-fired power generation. *J. Environ. Radioact.* 2, 145-160 (1985).

[26] Nakaoka, A., Fukushima, M., Takagi, S. Environmental effects of natural radionuclides from coal-fired power plants. *Health Phys.* 47, 407-416 (1984).

[27] Tomcynska, J., Blaton-Albicka, K., Pensko, J., Fugiel, D. The results of measurements of the natural radionuclides in coal power plants wastes and light concrete samples. Report TB-APF 80. Radiation Protection Department, Institute of Nuclear Research, Swierk, Poland, pp 9. (1980).

[28] Coles, D. G., Ragiani, R. C., Ondov, J. M. Behaviour of natural radionuclides in western coal-fired power plants. *Environ. Sci. Technol.* 12 (4), 442-446 (1978).

[29] Persson B. R. R., Holm E. Polonium-210 and lead-210 in the terrestrial environment: a historical review. *J. Environ. Radioact.*, 102, 420-429 (2011).

[30] Hamilton, E. I. The relative radioactivity of building materials. *Am. Ind. Hyg. Assoc. J.* 32, 398-403 (1971).

[31] Krisiuk, E. M., Tarasov, S. I., Shamov, V. P., Shalak, N. I., Lisachenko, E. P. and Gomelsky, L. G. A Study on Radioactivity in Building Materials. Research Institute for Radiation Hygiene (1971).

[32] Stranden, E. Some aspects on radioactivity of building materials. *Phys. Nor.* 8, 167-177 (1976).

[33] ICRP-60. Recommendations of the International Commmission on Radiological Protection. Pergmon Press (1990).

[34] UNSCEAR. Sources and effects of ionizing radiation. United Nations Scientific Committee on the Effects of Atomic Radiation (UNSCEAR) Report to the General Assembly with annexes, United Nations, New York (1993).

[35] Baeza A., Corbacho J. A., Gullén J., Salas A., Mora J. C. Analysis of the different source terms of natural radionuclides in a river affected by NORM (Naturally Occurring Radioactive Materials) activities. *Chemosphere* 83, 933-940 (2011).

[36] González-Labajo, J., Bolívar J.P., García-Tenorio R. Natural radioactivity in waters and sediments from a Spain mining river. *Radiat. Phys. Chem.* 61, 643-644 (2001).

[37] Lozano, J. C., Blanco, P., Vera Tomé, F. Distribution of long-lived radionuclides of the ^{238}U series in the sediments of a small river in a uranium mineralized region of Spain. *J. Environ. Radioact.* 63, 153-171 (2002).

[38] Flues M., Moraes V., Mazzilli B. P. The influence of a coal-fired power plant operation on radionuclide concentration in soil. *J. Environ. Radioact.* 63, 285-294 (2002).

[39] Dowdall M., Vicat K., Frearson I., Gerland S., Lind B., Shaw G. Assessment of the radiological impacts of historical coal mining operations on the environment of Ny-Alesund, Svalbard. *J. Environ. Radioact.* 71, 101-114 (2004).

[40] Gür F, Yaprak G. Natural radionuclide emission from coal-fired power plants in the southwestern of Turkey and the population exposure to external radiation in their vicinity. *J. Env. Sci. Health A Tox. Hazard Subst. Environ. Eng.* 45: 1900-1908 (2010).

[41] Nenadović S, Nenadović M, Kljajević L, Vukanac I, Poznanović M, Mihajlović-Radosavljević A, Pavlović V. Vertical distribution of natural radionuclides in soil: Assessment of external exposure of population in cultivated and undisturbed areas. *Sci. Total Environ.* in press (2012).

[42] Bem EM, Bem H, Wieczorkowski P. Studies of radionuclide concentrations in surface soil in and around fly ash disposal sites. *Sci. Total Environ.* 220: 215-222 (1998)
[43] Papastefanou C, Manolopoulou M, Charalambous S. Radioecological measurements in the coal power plant environment. *Radiation Prot. Dosimetry* 24: 439-443 (1988).
[44] Bunzl K, Hötzl H, Rosner G, Winkler R. Spatial distribution of radionuclides in soil around a coal-fired power plant: ^{210}Pb, ^{210}Po, ^{226}Ra, ^{232}Th, ^{40}K emitted with the fly ash and ^{137}Cs from the worldwide weapon testing fallout. *Sci. Total Environ.* 38: 15-31(1984).
[45] Rosner G, Bunzl K, Hötzl H, Winkler R. Low level measurements of natural radionuclides in soil samples around a coal-fired power plant. *Nucl. Instrum. Methods Phys. Res*. 223: 585-589 (1984).
[46] Sánchez-Cabeza J. A., García-Talavera M., Costa E., Peña, V., García-Orellana, J., Masqué, P., Nalda, J.C. Regional calibration of erosion radiotracers (^{210}Pb and ^{137}Cs). Atmospheric fluxes to soils (Northern Spain). *Environ. Sci. Technol.* 41, 1324-1330 (2007).
[47] Eisenbud M, Petrow HG. Radioactivity in the atmospheric effluents of power plants that use fossil fuels. *Science* 144: 288-289 (1964).
[48] Tzortzis M, Tsertos H. Determination of thorium, uranium and potassium elemental concentrations in surface soils in Cyprus. *J. Environ. Radioact.* 77: 325-338 (2004).
[49] Dugalic G, Krstic D, Jelic M, Nikezic D, Milenkovic B, Pucarevic M, Zeremski-Skoric T. Heavy metals organics and radioactivity in soil of western Serbia. *J. Hazard Mater.*; 177: 697-702 (2010).

In: Thorium
Editor: Renae Jamison

ISBN: 978-1-63321-309-8

Chapter 4

BEHAVIOR OF Th-ISOTOPES ALONG THE PIGMENT T_IO_2 INDUSTRIAL PRODUCTION PROCESS

M. J. Gázquez [1,3], J. Mantero [2], S. Pérez-Moreno [1], R. García–Tenorio [2] and J. P. Bolívar [1,*]

[1]Departamento de Física Aplicada, Universidad de Huelva, Huelva, España

[2] Departamento de Física Aplicada II Universidad de Sevilla, Sevilla, España

[3]Departamento de Química, Universidad Técnica Particular de Loja (UTPL), Loja, Ecuador

ABSTRACT

The present study, conducted initially to characterize from radiological point of view, the raw materials (ilmenite and titanium slag), co-products (copperas and sulfafer), wastes (red gypsum and ilmenite mud) and final product (TiO_2) used/generated along the titanium dioxide industry, located in the province of Huelva, has allowed in addition a detailed study of the routes followed by several radionuclides, with special emphasis on Th-isotopes, throughout the industrial process.

---

* Corresponding authors: bolivar@uhu.es.

This pigment TiO_2 industry works by the "sulphate method", which is a typical NORM (Naturally Occurring Radioactive Materials) industry. The raw material (ilmenite, mainly formed by $FeTiO_3$), contains around 100 Bq kg^{-1} of U-series radionuclides, and around 400 Bq kg^{-1} from the Th-series ones.

Along the industrial process, the secular equilibrium for the uranium and thorium series existing in the raw material is broken, and then the radionuclides are distributed/accumulated in the different materials according to their chemical behaviors. Therefore, the main goal of this study has been to gain basic information of the behavior of Th-isotopes in comparison with the behaviour of other natural radionuclides along the industrial process and, with basis in the information gained, to analyze the future potential environmental impact of co-products and wastes obtained in the industrial process. In this sense, it is important to note that some of the co-products (copperas and sulfafer) are currently used in fields such as, agriculture, fertilizers, etc.

Keywords: Titanium dioxide, thorium-isotopes, NORM, radioactivity

1. Introduction

In opposition of the popular belief, the most widely used titanium product is not the titanium metal or some of its alloys, but rather is the titanium dioxide (TiO_2) pigment that provide whiteness and opacity to a vast range of everyday products from coatings and plastics, to inks and even as flux in glass manufacture, filler in paper, cosmetic, food and rubber industries (Zhang et al, 2011). The production process of titanium dioxide, the most commonly used white pigment, usually involves one of two alternative processes, called "chloride" and "sulphate" routes (Langmesser et al., 1973; McNulty, 2007).

Located in the province of Huelva (southwest of Spain), and as a part of a big chemical industry complex, it can be found the only industry devoted to the production of titanium dioxide pigments in Spain.

The titanium dioxide production process in the Huelva´s plant is based in the acid digestion and dissolution of the feedstock (mix of 85% of ilmenite ($FeTiO_3$) with about 15% of titaniferous slag) by using sulphuric acid (98 %) (Gázquez et al., 2009). In addition, this type of industry is considered a NORM (Naturally Occurring Radioactive material) industry (IAEA, 2006), because of the raw material ($FeTiO_3$) contains around 100 Bq kg^{-1} of radionuclides from the uranium series and 400 Bq kg^{-1} from thorium-series radionuclides. Along the industrial process, the secular equilibrium between the radionuclides from the U and Th series is disrupted and these radionuclides are distributed/accumulated in the different materials according to the chemical behavior of each radioelement (Mantero et al., 2013).

For this reason, the main objective of this work has been to obtain information about the chemical behaviour of Th-isotopes along the most significant steps of the industrial process for TiO_2 pigment manufacturing, being compared this behaviour with the observed ones for other natural radionuclides (uranium and radium isotopes).

This has allowed a simple but effective assessment of the possible public radiological impact associated with the use of the products and co-products obtained in this type of industry as well as of the possible environmental radiological impact of the solid residues and liquid discharges generated. In this sense, it is important to note that the ferrous sulphates (hepta and monohydrated) obtained as a co-products, are currently used in the agriculture field, as supplier of iron to prevent chlorosis in plants

2. Industrial Process

In the Huelva TiO_2 industry, the oldest and most common process for titanium dioxide production called "sulphate process" is used. This process uses concentrated sulphuric acid (H_2SO_4) to dissolve the titaniferous feedstocks, ilmenite and slag. Ilmenite is a heavy mineral containing approximately 43% to 65% titanium dioxide (Chernet, 1999; Gázquez et al, 2009), while the slag contains 70% to 80% titanium dioxide. The slag is obtained from the smelting of ilmenite in order to decrease the reagent consumption and waste material accumulation in the subsequent pigment-making process (Pistorius and Coetzee, 2003).

A detailed diagram of the production process at the Huelva titanium dioxide factory is shown in Figure 1. This diagram also shows schematically the treatments applied to the two main co-products generated during the process. The main steps in the TiO_2 production process at the Huelva factory are the following:

Step A. A carefully controlled blend of illmenite and slag is mixed with highly-concentrated sulphuric acid (98%) and acid (H_2SO_4 80%) recycled from the process, in order to digest the TiO_2-containing feedstock. A highly-exothermic reaction is initiated by the addition of defined amounts of steam, water and diluted acid. The equation for the digestion reaction is summarized as:

$$FeTiO_3 + 2H_2SO_4 \rightarrow TiOSO_4 + 2H_2O + FeSO_4$$
$$TiOSO_4 + H_2O \rightarrow TiO_2n.H_2O + H_2SO_4$$

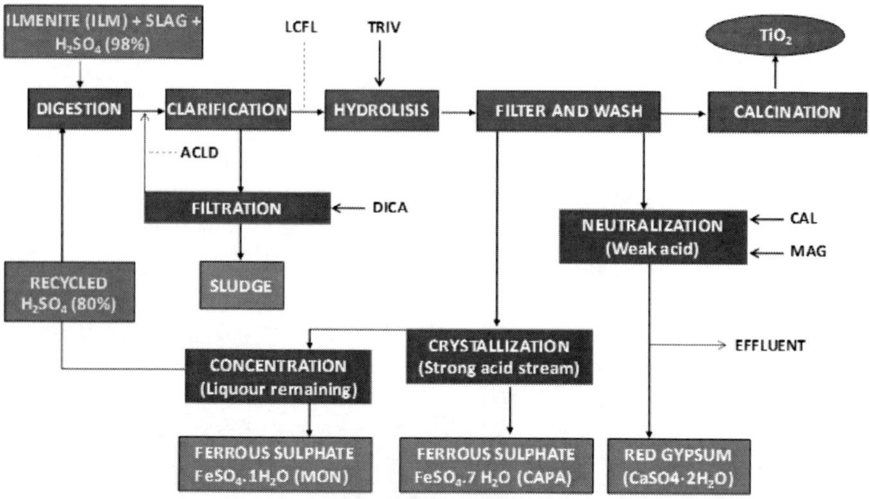

Figure 1. Detailed diagram of the production process.

The resulting liquor contains titanyl sulphate ($TiOSO_4$) and iron sulphate ($FeSO_4$) dissolved in sulphuric acid.

Step B. Clarification of the resulting liquor. The reduced liquor passes to a clarification tank where the undissolved solids (sludge) are allowed to settle; afterwards, they are separated from the solution by flocculation (decantation) and filtration. Nowadays, this sludge is neutralised and, finally, disposed of in a controlled disposal area.

Step C: Titanium dioxide precipitation, lixiviation and washing. The clarified liquor is hydrolyzed with steam in order to produce the precipitation of hydrated titanium. The TiO_2 precipitate is then separated from the mother liquor and submitted to a lixiviation process with sulphuric acid. After the lixiviation, the TiO_2 cake is separated again and washed with water in order to remove remaining impurities. The liquor resulting from the separation of the TiO_2 after hydrolysis and lixiviation form the first by-product of the process (strong acid, rich in Fe). The weak acid solution formed in the final wash is the second by-product of the process. The TiO_2 pulp obtained after washing is placed in rotary kilns for the removal of its water content and some traces of sulphur. The resulting solid is cooled, milled, coated, washed, dried and finely micronized, before be packed for commercial distribution.

Step D._Treatment of the strong acid. Initially, the first by-product, i.e. the mother liquor ("strong" acid, 20%-25% H_2SO_4) resulting from separation and lixiviation of the precipitated $TiO_{2,}$ is pumped to batch cooler crystallisers, where the bulk of the iron sulphate is removed as solid ferrous sulphate heptahydrate. This constitutes the first co-product of the TiO_2 process, commonly known as copperas (CAP). The remaining strong acid is then reconcentrated for reuse in the initial digestion step. This reconcentration is achieved by multi-stage evaporation to concentrate the acid, with the remaining ferrous sulphate in the reconcentrated acid solution being present at the high temperatures employed in the monohydrate rather than the heptahydrate form. This ferrous sulphate monohydrate (MON) precipitates in the reconcentration step, forming a solid co-product which is separated by filtration.

Step E. Treatment of weak acid. The second by-product, formed by the solution resulting from the final stages of washing $TiO_{2,}$ is also processed by sending it to a neutralization plant. The neutralisation step consists of adding lime, or limestone, to the weak acid stream, generating a co-product called red gypsum (RG), formed mainly of gypsum and iron hydroxides, according to the following reactions:

$$Ca(OH)_2 + H_2SO_4 \rightarrow CaSO_4.2H_2O$$
$$FeSO_4 + Ca(OH)_2 \rightarrow Fe(OH)_2 + CaSO_4$$

The red gypsum is separated by filtration, while the resulting clean water is partially recycled in the process with the remaining water is released into the estuary where the factory is located. The concentrations of pollutants in these waters are substantially lower than the limits fixed by the Spanish government for environmental emissions.

The magnitude of the co-products generated at the Huelva factory can be seen clearly in the following figures: Annually, around 142,000 metric tonnes of raw material are processed (80 % ilmenite and 20 % slag), with the resulting generation of 30,000 metric tonnes of sludge, 70,000 metric tonnes of RG, 140,000 metric tonnes of CAP and 125,000 metric tonnes of MON.

3. Material and Methods

3.1. Sample Collection and Pretreatment

Samples of the raw-materials, co-products, final product, residues and effluents have been collected. The collection was performed by personnel of the factory following the protocol adapted and refined by them in order to obtain representative samples for controlling the production process. The amount of material collected in each sampling point was about 1 kg for a wet solid sample, and 5 L for liquid, liquor material or pulps. The commented representativeness of the collected samples have been verified taking at each collection point and different times along a month, five additional samples, and checking the uniformity in their radioactive content (variations lower than 20% for all the radionuclides were observed in all the collection points).

A description of the samples is compiled in Table 1. These samples have been mainly classified in five groups according to the step of the process from which they were collected: (A) Digestion of the ore; (B) Clarification of the resulting liquor; (C) Titanium dioxide precipitation, lixiviation and washing, (D) Treatment and recycling of the strong acid, and (E) Treatment of the weak acid. The location of the collection for most of the samples is also indicated in Figure 1.

After collection, solid samples were dried at 105°C until reaching constant weight before analysis, except the ferrous sulphates (MON and CAP) and red gypsum that were dried at 45°C to avoid the loss of their hydration water.

The solid samples were afterwards homogenized mechanically, grinded and passed through a 0.8 mm sieve. In the case of liquid samples, they were gently heated before each analysis or before to take it a representative aliquot.

3.3. Radionuclide Determinations

3.3.1. Gamma-Ray Spectrometry

Aliquots of the pre-treated samples were introduced into polypropylene cylindrical containers with a diameter of 6.6 cm, up to a height of 5 cm for gamma-ray spectrometry. Samples were measured in an extended range Germanium coaxial detector (XtRa) of 37.1% relative efficiency with a 10 cm passive shielding of ancient lead (6 Bq/kg of ^{210}Pb) (Hurtado et al., 2003), and with an active shielding based in an organic scintillation detector working in coincident mode with the Ge detector.

Table 1. Codes and description of the collected and analyzed samples

STEP	CODE	DESCRIPTION
Digestion of the ore		
A	ILM	Ilmenite mineral. Imported raw material of the process
A	SLAG	Raw material from smelting of ilmenite
A	H_2SO_4·98%	Sulphuric acid.
A	H_2SO_4·80%	Recycled sulphuric acid from next step of the TiO2 process
Clarification of the resulting liquor:		
B	SLUDGE	Mud formed mainly by un-dissolved material separated after the digestion by decantation and filtration
B	DICA	Raw material used as a basis of the filters used for the separation of the sludge
B	ACLD	Liquor obtained by decantation and filtration of sludge
B	LCFL	Liquor enriched in Ti and Fe, and free of the un-attacked material.Collected before starting the Ti separation
Titanium dioxide precipitation, lixiviation and washing		
C	TRIV	Trivalent Ti added to induce the removal of Fe traces in the lixiviation step
C	TiO_2 pulp	Solid pulp collected after precipitation, lixiviation and washing
Treatment of strong acid		
D	CAP	Copperas. Iron sulphate heptahydrated generated by crystallization
D	MON	Iron sulphate monohydrate. Precipitate obtained after the concentration of the liquor remaining after the precipitation of the copperas.
Treatment of weak acid		
E	CAL	Ca hydroxide, coming from outside the process, and used for the neutralization of the acidic waters
E	MAG	Mg hydroxide coming from outside the process, and used for the neutralization of the acidic waters
E	RED GYP	Precipitate generated in the neutralization process
E	EFFLUENT	Waters with origin in the filtration of the precipitate formed in the neutralization process, after its decantation.

For gamma-ray quantitative determinations, we have applied an original efficiency calibration procedure in the energy range of 150–1800 keV, taking into account the self-absorption differences between the problem sample and the calibration sample (Perez-Moreno et al., 2002). ^{226}Ra activities were determined by the 352 keV emission of ^{214}Pb. Both nuclides were in secular equilibrium since counting was done at least one month after filling and sealing the cylindrical container. ^{228}Ra determination was carried out by the emission of 911 keV of ^{228}Ac.

3.3.2. Alpha-Particle Spectrometry

Aliquots of each sample were taken for the determination of several alpha-emitters by alpha-particle spectrometry. U-isotopes (^{238}U, ^{235}U, ^{234}U) and Th-isotopes (^{232}Th, ^{230}Th) activity concentrations by alpha-particle spectrometry were determined. To this end, a sequential well-established radiochemical method was applied using extraction chromatographic resins (Pilvio and Bickel, 2000) and using ^{232}U and ^{229}Th as internal tracers. In the case of solid samples, the conventional radiochemical method for alpha spectrometry starts with the application of a wet-digestion with a mixture of concentrated HNO_3 and HCl in order to put in dissolution the radionuclides of interest. This wet-digestion has proven to be effective in most cases, but it cannot afford the total dissolution of the radionuclides when the sample being treated has a refractory fraction in its composition. For that reason, the two main raw materials (ilmenite and slag), and the undissolved mud (SLUDGE) formed in the digestion process have been submitted to a fusion method in order to achieve its total dissolution (Mantero et al., 2010). To the fused solution was then coupled the conventional method with extraction chromatographic resins in order to obtain the isolated U and Th sources.

In this method, the U-isotopes and the Th-isotopes once isolated are electrodeposited onto stainless steel planchets (Hallstadius, 1984). Typical recoveries of 70–90% for uranium and 50–70% for thorium were obtained. The activity uncertainties were calculated through error propagation of the physical variables (with coverage 1 sigma) while the typical MDA values for the alpha-emitters analyzed (Currie, 1968) are three orders of magnitude lower than those for the gamma-emitters.

4. Results and Discussion

Table 2 shows the activity concentrations (Bq/kg) of several natural radionuclides belonging to the uranium and thorium series, whose half-lives are longer than one month. As was mentioned above, ilmenite is the main raw material used in the titanium dioxide industrial process and shows a radioactive content that can be evaluated as moderate in comparison with the range of activity concentrations which can be found over the world associated with this mineral (IAEA, 2006). The treated ilmenite contains around 300 Bq kg^{-1} for ^{232}Th and its progeny, and around 100 Bq kg^{-1} for the radionuclides of the uranium-series (including ^{230}Th). In addition, it is possible to affirm for both series the existence of secular equilibrium between their members.

Table 2. Activity concentrations (Bq.kg^{-1})and several ratios for the radionuclides of the both uranium and thorium series and ^{40}K, determined in the samples collected in the steps A and B of the production process. H.R. Relative humidity (%)

Step		R.H.	^{238}U	^{230}Th	^{226}Ra(^{214}Pb)	^{232}Th	^{228}Th(^{212}Pb)	^{228}Ra(^{228}Ac)	^{40}K
A	ILMENITE	3.1	119 ± 3	85 ± 5	92 ± 5	315 ± 20	305 ±23	301 ± 20	20 ± 2
A	SLAG	2.3	5.9 ± 0.7	14 ± 3	7 ± 2	19 ± 4	9.8 ±1. 7	9.0 ± 1.2	8 ± 2
A	H_2SO_4 98%	-	0.11 ± 0.03	1.0 ± 0.1	MDA	2.7 ± 0.2	MDA	MDA	MDA
A	H_2SO_4 80%	-	5.4 ± 0.3	176 ± 8	MDA	49 ± 3	164 ± 10	13 ± 1	MDA
B	SLUDGE	42	210 ± 30	45 ± 3	860 ± 30	350 ± 30	680 ± 30	2590 ± 60	284 ±11
B	DICA	15	1.3 ± 0.3	9.6 ± 1.2	64 ± 6	5.0 ± 0.8	129±17	91 ± 7	1124 ± 77
B	ACLD	-	7.8 ± 0.5	19 ± 1	MDA	62 ± 3	49 ± 6	6.6 ± 1.0	MDA
B	LCFL	-	12 ± 1	34 ± 2	1.2 ± 0.5	120 ± 5	72±9	16 ± 1	MDA
C	TRIV	-	0.5 ± 0.1	1.5 ± 0.2	MDA	0.4 ± 0.1	MDA	MDA	MDA
C	TiO2 PULP	42	0.8 ± 0.3	0.9 ± 0.6	3.0 ± 0.4	1.8 ± 0.3	3.8 ± 0.5	12.4 ± 1.8	MDA
D	COPPERAS	35	1.4 ± 0.2	2.8 ± 1.5	MDA	12 ± 1	18 ± 2	4 ± 1	MDA
D	MON	7.2	71 ± 2	113 ± 7	8 ± 1	500 ± 27	497 ± 37	60 ± 6	MDA
E	CAL	2.6	3.7 ± 0.2	6.9 ± 0.4	5 ± 1	0.9 ± 0.1	MDA	MDA	49 ± 5
E	MAG	8.2	7.1 ± 0.4	8.9 ± 0.4	10.5 ± 1.1	5.3 ± 0.4	5.5 ± 0.6	4.7 ± 0.5	26 ± 3
E	RED GYP	40	15 ± 1	41 ± 2	13 ± 2	143 ± 5	93 ± 6	69 ± 5	24 ± 9
E	EFFLUENT	-	0.004 ± 0.002	0.05 ± 0.01	MDA	0.06 ± 0.01	MDA	MDA	MDA

Moreover, the slag used as feedstock in the process, mixed with ilmenite in a proportion 80 % ilmenite and 20 % slag, contains quite low activity concentrations for all natural radionuclides and are lower than those found in typical undisturbed soil (35 and 30 Bq/kg for ^{238}U and ^{232}Th, respectively) (UNSCEAR 2000). If we remember that the slag is formed by submitting to ilmenite to a smelting process for diminishing the amount of Fe entering in the TiO_2 production process, it is clear to conclude that most of the radionuclides originally present in the ilmenite do not remain in the slag, accompanying then to the Fe and impurities removed. This fact is important from a radiological point of view, because an increase in slag proportion used in the raw material will reduce the possible radiological impact generated by the TiO_2 industry.

It is also quite interesting to comment the results obtained in the analysis of the sulphuric acid samples collected. The strong sulphuric acid (98%), comes from outside the TiO_2 process, and it is characterized for presenting a negligible radioactive content. On the other hand, the recycled sulphuric acid (80%), presents high activity concentrations of Th-isotopes, in minor extent of isotopes of uranium, and negligible concentrations of radium isotopes, fact that is directly related with the different routes followed by the different elements along the process and with the origin of this recycled acid, as we can see later. The fractionation of radionuclides is evaluated through their activity ratios (Table 3). In this table the ^{230}Th/^{238}U activity ratio present values up to around 30 in the recycled sulphuric acid, demonstrating this result that ^{230}Th tends to be accumulated in this acid. In addition, the ^{230}Th/^{232}Th activity ratios in raw material and recycled sulphuric acid are very different, showing this result that Th and U has to be bound to different minerals with different solubility in sulphuric acid. The different behavior in acid media of ^{228}Th and ^{232}Th, isotopes of the same radioelement, with an activity ratio upper than 3, could be explained by the fact that ^{228}Th comes from an alpha decay of ^{232}Th, and this has generated a preferential leaching of its daughter, the ^{228}Th.

Table 3. Activity ratios more significant of the industrial process

Step		^{228}Ra/^{232}Th	^{228}Ra/^{226}Ra	^{230}Th/^{238}U	^{228}Th/^{232}Th	^{232}Th/^{230}Th
A	ILMENITE	0.96 ± 0.09	3.3 ± 0.3	0.71 ± 0.05	0.97±0.10	3.7±0.3
A	SLAG	0.47 ± 0.12	1.3 ± 0.4	2.4 ± 0.6	0.52±0.14	1.4±0.4
A	H_2SO_4 98%	-	-	9.1 ± 2.6	-	2.7±0.3
A	H_2SO_4 80%	0.27 ± 0.03	-	33 ± 2	3.3±0.3	0.28±0.02
B	SLUDGE	7 ± 1	3.0 ± 0.1	0.21 ± 0.03	1.9±0.2	7.8±0.8
B	DICA	18 ± 3	1.4 ± 0.2	7.4 ± 1.9	26±5	0.52±0.11
B	ACLD	0.11 ± 0.02	-	2.4 ± 0.2	0.79±0.10	3.3±0.2
B	LCFL	0.13 ± 0.01	13 ± 6	2.8 ± 0.3	0.60±0.08	3.5±0.3
C	TRIV	-	-	3.0 ± 0.7	-	0.27±0.08
C	TiO2 PULP	6.9 ± 1.5	4.1 ± 0.8	1.1 ± 0.9	2.1±0.4	2.0±1.4
D	COPPERAS	0.33 ± 0.09	-	2.0 ± 1.1	1.5±0.2	4.3±2.3
D	MON	0.12 ± 0.01	7.5 ± 1.2	1.6 ± 0.1	1.0±0.1	4.4±0.4
E	CAL	-	-	1.9 ± 0.1	-	0.13±0.02
E	MAG	0.89 ± 0.12	0.45 ± 0.07	1.3 ± 0.1	1.0±0.1	0.60±0.05
E	RED GYP	0.48 ± 0.04	5.3 ± 0.9	2.7 ± 0.2	0.65±0.05	3.5±0.2
E	EFFLUENT	-	-	13 ± 7	-	1.2±0.3

In relation to the results obtained in the sample collected along the clarification step, it is interesting to comment that the sludge, formed mainly by un-dissolved raw material in the digestion of the feedstock, is highly enriched in Ra-isotopes. In this sense, we can affirm that this material should be considered and treated as a NORM waste. The sludge contains a total concentration of radionuclides greater than 1 Bq/g (IAEA, 2004), which is a considerable fraction of the radioactivity content originally present in the raw materials. The activity concentrations of ^{226}Ra and ^{228}Ra in the sludge are higher than the activity concentrations found in the ilmenite but with the same $^{228}Ra/^{226}Ra$ activity ratios, around 3, see table 3. The association of the great majority of Ra with the mud has a straightforward explanation; in the digestion of the ilmenite with the concentrated sulphuric acid, the radium presents a sulphate form, being well known the insolubility of the radium sulphate in acidic media (Landa E.R., 2007).

The results obtained in the sample of mud (sludge) indicate that the majority of the U and Th originally present in the raw material is dissolved in the digestion step and flow into the following steps of the process. Indeed, for example, for ^{232}Th the activity concentration determined in the mud sample (350 $Bqkg^{-1}$) is reflecting that only 10% of this element remains associated with the undissolved material, since for each mass unity of ilmenite 0.2 units of wet mud is generated (0.10 g in dry form) (Gazquez et al, 2011). In order to provide a full explanation of the results compiled in Table 2 for the sludge samples, it is necessary to point out that the secular equilibrium, which originally is given in the mineral for both U- and Th-series, is broken when is chemically treated with concentrated sulphuric acid. This explains, for example, why the concentration of ^{228}Th (680 Bq/kg, and half life = 1.9 y) is 2 times higher than ^{232}Th (350 Bq/kg) in the mud samples, even though ^{232}Th and ^{228}Th are isotopes of the same element (and therefore have the same behaviour during the industrial process), belong to the same radioactive series, and for that they will be in secular equilibrium in the raw material. The difference in the levels is due to the high initial concentration of ^{228}Ra in the mud in relation to the ^{228}Th one. This ^{228}Ra decays over time and increases the concentration of ^{228}Th activity over the concentration of ^{232}Th, reaching the secular equilibrium ^{228}Th-^{228}Ra in around 6-8 years.

In the process of precipitation by hydrolysis, and after the lixiviation and washing (step C), the TiO_2 pulp recovered is practically free of radioactivity. This fact indicates that the lixiviation and washing step is extremely effective in the removal of uranium and thorium in the precipitated pulp. In addition, it is interesting to highlight the fact that on the frame of the very low activity concentrations found in the TiO_2 pulp after lixiviation and washing, the higher concentrations correspond to ^{228}Ra, with a $^{228}Ra/^{226}Ra$ activity ratio compatible with the activity ratio shows in the raw material (ilmenite), table 3. In conclusion, both uranium and thorium isotopes have not a high tendency to be present in the TiO_2 final sample. The TiO_2 pulp is afterwards calcined, milled, coated, washed, dried and finely ground ("micronized"), before be packed for commercial distribution. It can be affirmed then in view of the obtained results that the main commercial product is practically free of radioactivity, in agreement with the results obtained in other studies (McNulty, 2008).

In the process for obtaining the clean TiO_2 pulp, two main liquid solutions (strong acids and acid waters) are generated. The activity concentrations obtained in samples collected in these steps of the process (Step D: *Treatment of the strong acid* and Step E: *Treatment of the weak acid*) are compiled in Table 2.

In the step devoted to the treatment of the strong acid remaining after the precipitation and lixiviation of the pulp, firstly it can be observed that in the process of crystallization, the material formed and precipitated (copperas) is practically free of radioactivity in relation with an undisturbed soil [31], being the $^{232}Th/^{230}Th$ activity ratio similar to obtained for the ilmenite, table 3. The crystals formed in this step are quite pure not containing in its structure as impurities some of the radionuclides originally present in the acid submitted to crystallization.

The liquor remaining after the crystallization step is then submitted to concentration and maduration in two steps, until the ferrous sulphate monohydrate is formed and precipitated (sample MON) and a relatively clean sulphuric acid solution is obtained. In this step, most of the radionuclides remaining in the treated liquor precipitate with the ferrous sulphate monohydrate, giving particularly activity concentrations of uranium and thorium isotopes in this co-product of the same order that the determined ones in the ilmenite. Thus, for example 113 Bq kg^{-1} for ^{230}Th and around 500 Bq kg-1 for ^{228}Th and ^{232}Th are measured in the MON sample. Therefore, the majority of thorium dissolved from the raw material in the digestion step flows then along the process until it co-precipitates in the monohydrate ($FeSO_4 \cdot H_2O$).

The relatively clean sulphuric acid, remaining after the precipitation of the monohydrate, is concentrated by evaporation until 80%, and recycled in the process to be introduced into the digestion step. This recycled acid contains not negligible amounts of thorium isotopes (as it was commented previously and it can be observed in the results compiled in Table 2.

In Table 2 are also included the results obtained in the analysis of the samples collected in the step E of the process (*Treatment of the weak acid*). The obtained results indicate that the process of neutralization applied is very effective from the radioactive point of view, being the majority of the radionuclides contained in the weak acid bound to the red-gypsum. The final clean liquid effluent is discharged into the estuary of Huelva being practically free of natural radionuclides. In the red gypsum only the radionuclides of Th-isotopes appear with concentration higher than a typical undisturbed soils; 143 and 93 Bq kg^{-1} for ^{232}Th and ^{228}Th, respectively. In the final effluent the activity concentrations of all the radionuclides measured by gamma-ray spectrometry are below the minimum detectable activity.

The solid formed in the neutralization process (RED GYP), although accumulates most of the radioactive content originally associated with the weak acid, contains moderate radioactivity concentrations. The radionuclide content in the final effluent (EFFLUENT) is very small when compared with the gypsum precipitated formed from calcium and magnesium hydroxides products. Moreover these products contain very low activity concentrations in radionuclides from the uranium and thorium series (4 Bq/kg of ^{238}U and 1Bq/kg of ^{232}Th in the case of the calcium hydroxide, and 7 Bq/kg of ^{238}U and 5 Bq/kg of ^{232}Th in the case of magnesium hydroxide). Taking in consideration that for each gram of ilmenite treated, a total of 0.6 grams of wet red-gypsum are generated (0.33 g in dry form), we can affirm that only a small fraction of the natural radionuclides originally present in the ilmenite finish accumulated in this by-product.

In table 4 we can see the concentration factor for Th-isotopes, defined as the ratio between of the sample concentration and in the raw material. It is evident that only for ferrous sulphate monohydrate sample this factor is upper than 1 for all Th-isotopes. For recycled acid (H2SO4 80%) appear a concentration factor for ^{230}Th around 2, and in the sludge the ^{228}Th presents a similar factor, directly related to the high activity concentration of ^{228}Ra as we have explained previously.

In the light of the above comments, after the study of the whole process, we can ensure that the environmental radioactive impact of the activities performed by this NORM industry can be evaluated as negligible. A series of processes, devoted to the generation of co-products and to the recycling of the strong acids used along the process on one hand, and to the neutralization of the acid waters on the other hand, implies that only controlled releases of water, practically free of radioactivity, are performed to the surrounding environment. The only solid waste generated (sludge) contains significant amounts of radionuclides (especially for ^{228}Ra, about 2600 Bq kg^{-1}) and it is managed in such a way that its possible environmental and public impact is kept to a minimum.

The possible radiological impact of the commercialized products can be also evaluated as minimum.

Table 4. Concentration factor (ratio of the concentration of a sample between the raw material one) for ^{232}Th, ^{230}Th and ^{228}Th

	232Th			*230Th*			*228Th*		
ILMENITE	1.0	±	0.1	1.0	±	0.1	1.0	±	0.1
SLAG	0.06	±	0.01	0.16	±	0.04	0.03	±	0.01
H_2SO_4 98%	0.009	±	0.001	0.012	±	0.001		-	
H_2SO_4 80%	0.16	±	0.01	2.1	±	0.2	0.54	±	0.05
SLUDGE	1.1	±	0.1	0.53	±	0.05	2.2	±	0.2
DICA	0.016	±	0.003	0.11	±	0.02	0.42	±	0.06
ACLD	0.20	±	0.02	0.22	±	0.02	0.16	±	0.02
LCFL	0.38	±	0.03	0.40	±	0.03	0.24	±	0.03
TRIV	0.0013	±	0.0006	0.018	±	0.003		-	
TiO_2 PULP	0.006	±	0.002	0.011	±	0.007	0.012	±	0.002
COPPERAS	0.04	±	0.01	0.03	±	0.02	0.06	±	0.01
MON	1.6	±	0.1	1.3	±	0.1	1.6	±	0.2
CAL	0.003	±	0.001	0.08	±	0.01		-	
MAG	0.017	±	0.002	0.10	±	0.01	0.018	±	0.002
RED GYP	0.45	±	0.03	0.48	±	0.04	0.30	±	0.03
EFFLUENT	0.0002	±	0.0001	0.0006	±	0.0001		-	

The main commercial product, the TiO_2 pigments, used in coatings such as paints, plastics, high grade papers and printing inks is practically free of radioactivity, as in the case of one of the co-products formed, the copperas. This copperas is totally valorised by the Huelva company and used for the recovery of basic soils in agriculture, as a component of animal feed, in the production of cements and in the treatment of waters as primary flocculent.

The other main co-product generated, the ferrous sulphate mononohydrate (sample MON), has a well established market, being used as a raw material for the production of ferric fertilizers (used for the mitigation of the chlorosis phenomena in plants when they grow on soils deficient in iron), and in spite of its well characterized radionuclide content generates a minimum public and environmental radioactive impact, because this co-product is only enriched in uranium and thorium isotopes (majoritary alpha emitters) that provokes only minimum doses rates by external radiation. In addition, its use as fertilizer and its dilution in the fields treated with this co-product restrict its impact to a minimum (the phosphate fertilizers commonly used in agriculture contain comparable activity concentrations of uranium and thorium isotopes that the monohydrate, are used at similar rates, and it is well known that its radiological impact is minimum). Theoretically, the only possible way of radiological impact of the monohydrate, could be through inhalation by the workers handling it, but this pathway can be discarded if we take into account that this co-product is very hygroscopic, fact that minimizes the formation of aerosols from this material.

CONCLUSION

In this work has been developed a study trying to evaluate the behaviour of Th-isotopes along the industrial process of TiO_2 pigment production by the sulphate via. It has been demonstrated that Th-isotopes tend to be bound to the sulphuric acid fraction, which is recycled in the process, and therefore Th tends to be concentrated in the co-product called monohydrate ferrous sulphate ($H_2SO_4 \cdot H_2O$) presenting concentrations of about 500 Bq kg^{-1} and 100 Bq kg^{-1} for $^{232,228}Th$ and ^{230}Th, respectively. This co-product is recycled and valorised in several applications as fertilizers or cements production. Therefore, radiological assessments of the applications of this co-product should be developed, although it can be estimated in advance its minimum radiological impact. The another solid material presenting high radionuclide concentrations is the un-dissolved ilmenite (MUD sample), which present very high concentrations of Ra-isotopes, with concentrations of about 2600 and 900 Bq kg^{-1} for ^{228}Ra and ^{226}Ra, respectively, and for that this is a NORM waste.

The neutralization treatment applied to the final weak acid liquid effluent coming from the cleaning of TiO_2 pulp gives waters with a very good radiological quality, and for that they are directly discharged into the Estuary of Huelva, no producing a measurable environmental radiological impact in their surroundings. And as final conclusion, to point out that this industrial activity is a paradigmatic example how an adequate management of the by-products and effluents can end in obtaining co-products of commercial value with a defined market, minimizing the possible public and environmental radioactive impact of the NORM industry.

ACKNOWLEDGMENTS

This research has been partially supported by the Government of Andalusia´s project "Characterization and modelling of the phosphogypsum stacks from Huelva for their environmental management and control" (Ref.: RNM-6300) and by National Institution of Higher Education, Science, Technology and Innovation of the Republic of Ecuador -(SENESCYT for its acronym in Spanish), Prometeo Project. The authors would like to acknowledge the financial support received from the company Tioxide-Huelva by the research projects "Valorization of red gypsum from the industrial production of titanium dioxide" (PROFIT, CIT-310200-2007-47) and "Applications of red gypsum and Tionite waste in commercial applications".

REFERENCES

Chernet T. Applied mineralogical studies on Australian sand ilmenite concentrate with special reference to its behavior in the sulphate process. *Minerals Engineering*, Vol 12. No 5, 485-495, 1999.

Currie L.A,, 1968. Limits for Qualitative Detection and Quantitative Determination.

Gazquez M.J., Bolivar J.P., García-Tenorio R. and Vaca F., 2009. Physicochemical characterization of raw materials and co-products from the titanium dioxide industry. *J. Hazard. Mat.* 166, 1429-1440.

Gázquez M.J., Mantero J., J.P. Bolívar, García-Tenorio R., Vaca F., Lozano R.L. (2011) Physico-chemical and radioactive characterization of TiO_2 undissolved mud for its valorization. *J. Hazard. Mat.* 191, 269–276.

Hallstadius L., 1984. A method for the electrodeposition of actinides. *Nucl. Instrum. Meth.,* Section A 223, 266-267-

Hurtado S., García-Tenorio R. and García-León M., 2003. ^{210}Pb determination in lead shields for low-level gamma-spectrometry applying two independent radiometric techniques. *Nucl. Instrum. Meth.* Section A, 497, 381-388.

IAEA (International Atomic Energy Agency). Safety standards series, Application of the Concepts of Exclusion Exemption and Clearance, Safety guide No RS-G- 1.7. August 2004. STI/PUB/1202.

Landa E.R., 2007. Naturally occurring radionuclides from industrial sources: characteristics and fate in the environment. *Radioactivity in the Environment* 10, 211- 237.

Langmesser P.W., H.G. Volz, G. Kienast, (1973). Process leading to the production of titanium dioxide pigment with a high degree of whiteness, U.S. Patent 3,760,058.

Mantero J., Lehritane M., Hurtado S., García-Tenorio R., 2010. Radioanalytical determination of actinoids in refractory matrices by alkali fusion. *J. Radioanal. Nucl. Chem.,* 286, 557–563.

Mantero, J., Gazquez, M.J., Bolivar, J.P., Garcia-Tenorio, R., Vaca, F.2013. Radioactive characterization of the main materials involved in the titanium dioxide production process and their environmental radiological impact. *J. Environ Radioactiv.* 120, 26-32.

McNulty G.S., Production of titanium dioxide, in: Proceedings of NORM V International Conference, Seville, Spain, 2007, pp. 169–189.

Perez-Moreno J.P., SanMiguel E., Bolivar J.P. and Aguado J.L., 2002. A comprehensive calibration method of Ge detector for low level spectrometry measurement. *Nucl. Instrum. Meth.* Section A. 491, 152-162.

Pesl J., Eric, R.H., 2002. High temperature carbothermic reduction of Fe_2O_3–TiO_2–MxOy oxide mixtures. *Miner. Eng.* 15, 971– 984.

Pilvio R. and Bickel M., 2000. Actinoid separations by extraction chromatography. Appl. Radiat. Isot. 53, 273- 277.

Pistorius P.C., Coetzee C., 2003. Physicochemical aspects of titanium slag production and solidification. *Metall. Mater. Trans*., B 34B, 581– 588.

United Nations Scientific Committee On The Effects Of Atomic Radiation (Unscear) (2000). *Report of the United Nations Scientific Committee on the Effects of Atomic Radiation*, United Nations, New York.

Zhang, W; Zhu, Z & Cheng, C.Y. (2011). A literature review of titanium metallurgical processes. *Hidrometallurgy*, Vol 108; No 3-4; (July 2011), pp. 177-188. ISSN: 0304-386X.

In: Thorium
Editor: Renae Jamison

ISBN: 978-1-63321-309-8

Chapter 5

SOME FEATURES OF THE Th PRESENCE IN THE POÇOS DE CALDAS ALKALINE MASSIF, MINAS GERAIS STATE, BRAZIL

Daniel Marcos Bonotto [*]
Departamento de Petrologia e Metalogenia, Universidade Estadual Paulista (UNESP), Câmpus de Rio Claro, São Paulo, Brasil

ABSTRACT

Alkaline rocks are found in southern Brazil in association with the Ordovician-Cretaceous sedimentary Paraná basin and mainly distributed along its margins. They vary considerably in age and composition, and have been assembled into distinct geographic provinces, such as the Poços de Caldas alkaline massif, Minas Gerais State, representing one of their most important ones, particularly due to the presence of economical deposits bearing uranium, thorium and rare earth elements. Several rock types are associated with high ^{232}Th levels, mainly occurring at Morro do Ferro that is situated near the center of the Poços de Caldas plateau and initially drew the attention of mining prospectors due to the abundant occurrence of magnetite veins and dykes. It is a small hill located near the intersection of a fractures system, rising some 140 meters above the plateau base level and lying five kilometers to the north of the Osamu Utsumi uranium mine and about 10 km to the south of the city of Poços de Caldas. The superficial radioactivity in this area was systematically

[*] Corresponding author: Email: danielbonotto@yahoo.com.br.

investigated in the middle 60s with the aim of evaluating the effects of the accentuated radioactivity in the ore body environment. The level of the environmental gamma radiation close the top of the hill varied between 1 and 3 mR/h (100-300 times higher than the background) in an area of about 30,000 m^2 and the plants were so radioactive due to the ^{228}Ra absorption that could generate *autoradiographic images.* Thorium is a highly insoluble element in water due to its presence in refractory minerals of difficult dissolution, but depending on the pH, ionic strength and others factors it may be released to the liquid phase. This chapter reports the results of some studies focusing the thorium presence in the Poços de Caldas alkaline massif with the aim of addressing some insights on its behavior in the environment.

INTRODUCTION

Primeval radionuclides have survived in detectable amounts since the time of nucleosynthesis and contribute to the natural terrestrial radioactivity. The major contribution comes from ^{40}K and from radionuclides generated through the sequence of decay transformations of the three alpha emitting radionuclides: ^{232}Th, ^{238}U, and ^{235}U. Each radioactive series ends by a stable Pb isotope after passing through several unstable members having half-lives much shorter than that of the respective precursors. The members in each decay chain have atomic numbers between 81 and 92.

The natural radioelements uranium and thorium are lithophile elements widely distributed in crustal rocks, being concentrated preferentially in acid igneous rocks compared with intermediate, basic, and ultrabasic varities. Thorium is an element ~3-4 times more abundant than uranium in crustal rocks because it is less susceptible to mobilization in the supergene environment. It occurs predominantly as a tetravalent cation and a trace constituent in phosphates, simple and multiple oxides and silicates, as well in the major, rock-forming minerals such as monazite, thorianite (ThO_2) and thorite ($ThSiO_4$), among others [1,2]. The major natural Th isotope is ^{232}Th that is precursor of the mass number 4n decay series (Figure 1). The ^{232}Th half-life is 14.1×10^9 years and its decay series finishes at the stable ^{208}Pb (Figure 1). Table 1 reports some relevant data that are useful for the evaluation of radionuclides belonging to the ^{232}Th decay series. ^{232}Th is an emitter of alpha particles, where the stable ^{208}Pb in its decay series is attained after 12 disintegrations, i.e. 7 alpha-type and 5 beta-type (Figure 1). Guidance levels in drinking water for ^{232}Th and some its daughters like ^{228}Th and ^{228}Ra have been established by [3]. This chapter describes the presence of the thorium isotopes ^{232}Th and ^{228}Th in groundwater occurring at the Poços de Caldas alkaline massif, southeastern Brazil.

DETECTION OF ALPHA PARTICLES IN THE ^{232}TH DECAY SERIES

The alpha spectrometry is a useful technique often utilized to characterize the presence of natural thorium isotopes dissolved in water. It is based on the direct measurement of the alpha particles generated in the ^{232}Th decay series (Table 2). The Si(Au) surface barrier detectors have been widely used for characterizing the natural thorium isotopes, which are inserted within vacuum chambers in order to avoid absorption of alpha particles by the atmosphere. The alpha spectrometric systems are available for a variable number of Si(Au) surface barrier detectors, depletion depths and surface area in order to yield appropriate spectra for providing information about the activity concentration of dissolved thorium.

The α-spectra have been sometimes recorded on EG&G ORTEC 919 Spectrum Master Multichannel Buffer that provides four 1,024-channel analyzers controlled by MAESTRO software. Figure 2 illustrates a complete spectrometric system utilizing a detector coupled to a 1,024 multichannel analyzer. The system must be calibrated in energy on using different radioactive sources. One appropriate source has been prepared for this purpose at the *Centre de Faibles Radioactivités*, CNRS-CEA, Gif-sur-Yvette, France, and contains 0.91 Bq of ^{238}U and 0.17 Bq of ^{232}U that is an artificial uranium isotope produced in nuclear reactor by ^{232}Th neutron activation. It has a half-life of 72 years (Figure 3) and originates ^{228}Th by the emission of α-particles of 5.3 MeV energy (Figure 4) [7]. The ^{228}Th-daughters correspond to those occurring in the ^{232}Th decay series (Figures 1 and 3).

Before the use of alpha spectrometry, the dissolved thorium must be extracted from waters and deposited in metallic discs that are made of stainless steel, for instance, as described by [8]. The dissolved thorium present in aquifer systems has different chemical affinities and specific procedures should be utilized for performing its extraction from water samples before conducting the alpha-particles analysis. The dissolved thorium is generally found in very low concentrations and, therefore, it is a common practice to sample large (20-30 L) water volume that is stored in polyethylene bottles.

If ^{232}U-^{228}Th is utilized as spike for providing the evaluation of the chemical yield and thorium concentration, then, the volume sampled for ^{232}Th and ^{228}Th analyses should be divided into two aliquots of equal weight, with the ^{232}U-^{228}Th radioactive tracer being added to one aliquot. Both aliquots should be filtered through a 0.45 μm Millipore membrane and acidified to pH less than 2 using HCl, with addition of $FeCl_3$ to each aliquot.

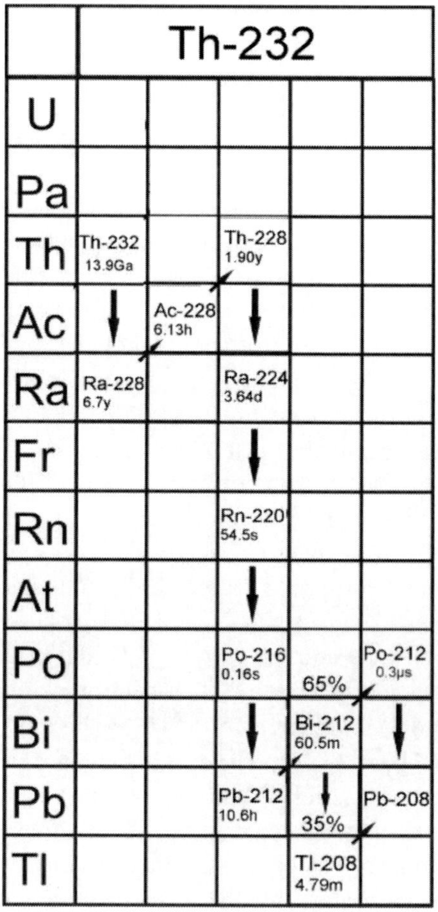

Figure 1. The thorium natural decay series. Modified from [4].

Table 1. Energy and electromagnetic transition related to the radionuclides decay belonging to the ^{232}Th decay series. According to [5]

Nuclide	Half life*	TD*	Particle energy/ Probability trans.		Electromagnetic transition					
			Energy (MeV)	Probabil. (%)	Energy (MeV)	F* (%)	Ti* (%)	Energy (MeV)	F* (%)	Ti* (%)
Th-232	1.405×10^{10} y	α	3.830	~ 0.2	0.059	0	22			
			3.953	~ 23	0.124	0	0.2			
			4.012	~ 77						
Ra-228	5.75 y	β^-	0.048	100	0.007	0	100			
Ac-228	6.13 h	β^-	0.45	4.9	0.058	0.6	79.4	0.783	1.4	
			0.49	5.6	0.099	0	5	0.796	4	
			0.62	5.7	0.129	2.8	7.2	0.836	2.5	
			0.99	7.3	0.184	1.7	5.6	0.912	23	
			1.02	4.0	0.209	4.3	0.2	0.966	7	
			1.12	6.5	0.270	4.1	0.2	0.970	13	
			1.17	33	0.328	5.3	0.2	1.464	1.1	
			1.76	20	0.338	15	1	1.503	2.5	
			2.10	13	0.410	2.5	0.5	1.593	4.5	
					0.463	3.6	0.6	1.642	2.4	
Th-228	1.913 y	α	5.140	0.03	0.085	1.6	27.4			
			5.176	0.2	0.132	0.19	0.01			
			5.211	0.4	0.167	0.12	0.18			
			5.341	28	0.216	0.29	0.01			

Table 1. (Continued)

Nuclide	Half life*	TD*	Particle energy/ Probability trans.		Electromagnetic transition					
			Energy (MeV)	Probabil. (%)	Energy (MeV)	F* (%)	Ti* (%)	Energy (MeV)	F* (%)	Ti* (%)
			5.424	71						
			others	low						
Ra-224	3.64 d	α	5.447	5.2	0.241	4.2	1.0			
			5.684	94.8						
Rn-220	55.3 s	α	5.747	0.07	0.542	0.07				
			6.288	99.93						
Po-216	0.15 s	α	5.984	~ 0.002	0.808	0.002				
			6.777	~100						
Pb-212	10.6 h	β^-	0.155	5	0.115	0.6	4.1			
			0.332	82	0.239	44.8	37.5			
			0.571	13	0.300	3.4	1.5			
Bi-212	60.6 m	α	5.607	0.4	0.040	1	28			
			5.768	0.6	0.288	0.3	0.1			
			6.051	25.2	0.328	0.1				
			6.090	9.6	0.453	0.4				
			others	low	0.727	6.6	0.1			
					0.785	1.1	0.1			
		β^-	0.445	0.7	0.893	0.4				
			0.572	0.3	0.952	0.2				

Nuclide	Half life*	TD*	Particle energy/ Probability trans.		Electromagnetic transition					
			Energy (MeV)	Probabil. (%)	Energy (MeV)	F* (%)	Ti* (%)	Energy (MeV)	F* (%)	Ti* (%)
			0.630	1.9	1.079	0.5				
			0.738	1.5	1.513	0.3				
			1.524	4.5	1.621	1.5				
			2.251	55.2	1.680	0.1				
			others	low	1.606	0.1				
Po-212	3.05×10^{-7} s	α	8.785	100						
Tl-208	3.07 m	β^-	1.032	3.4	0.253	0.8	0.6	0.860	12.3	0.3
			1.073	0.6	0.277	6.9	3.7	1.093	0.4	
			1.285	24	0.511	23.0	2.0	2.615	99.8	0.2
			1.518	22	0.583	85.8	2.2			
			1.795	50	0.763	1.8	0.1			
Pb-208	stable									

* y= years; d = days; m = minutes; s = seconds; F = Photons emitted; TD = Decay mode; Ti = Transitions internally converted.

Table 2. Alpha particles generated in the ^{232}Th decay series. Data reported by [6]

^{232}Th decay series		
Nuclide	**Energy (MeV)**	**Absolute Intensity (%)**
^{232}Th	3.95	24.0
	4.01	76.0
^{228}Th	5.34	28.0
	5.43	71.0
^{224}Ra	5.45	6.0
	5.68	94.0
^{220}Rn	6.29	100.0
^{216}Po	6.78	100.0
^{212}Bi	6.05	23.5
	6.09	9.0
^{212}Po	8.78	100.0

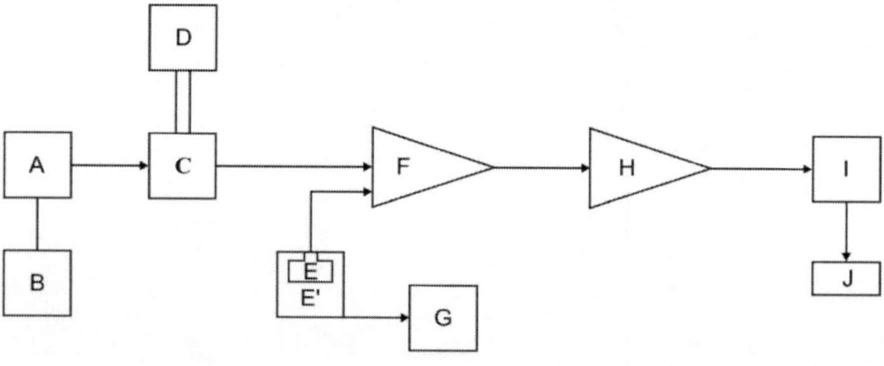

A - Power Supply - DC ± 18 V
B - Voltage Regulator - 110 V
C - Variable DC Power Supply (0-200 V)
D - Digital Voltmeter
E - Semiconductor Detector
E' - Vacuum Chamber
F - Pre-Amplifier
G - Vacuum Pump
H - Amplifier
I - Multichannel Analyzer
J - Recorder

Figure 2. A typical spectrometric system for detecting alpha particles.

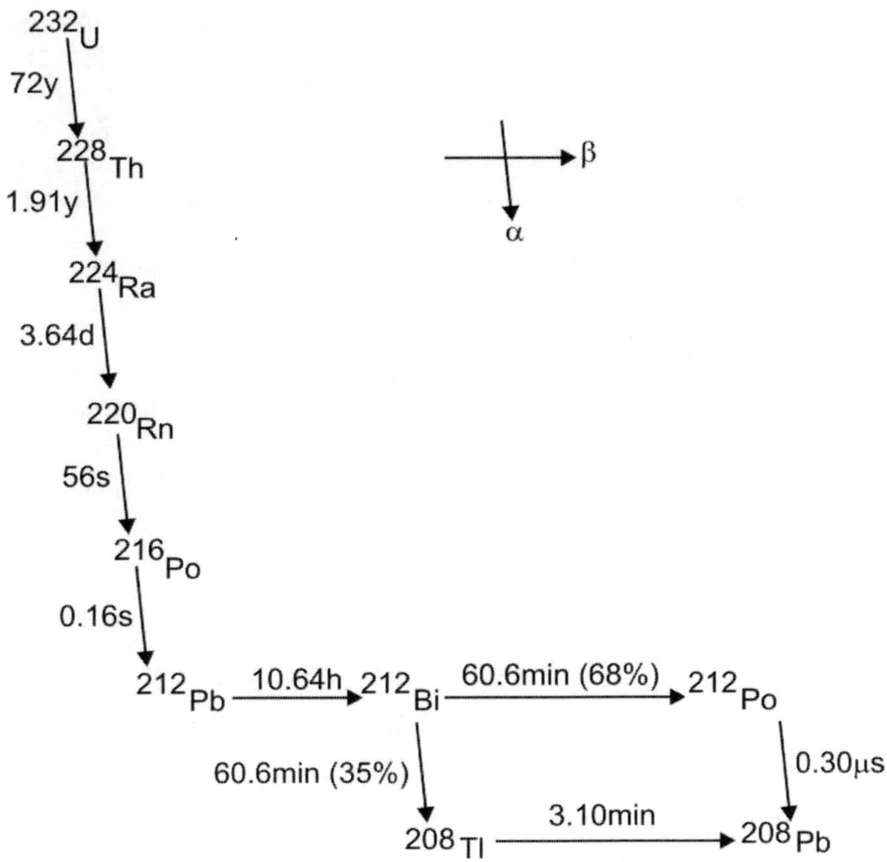

Figure 3. The ^{232}U decay chain. Modified from [9].

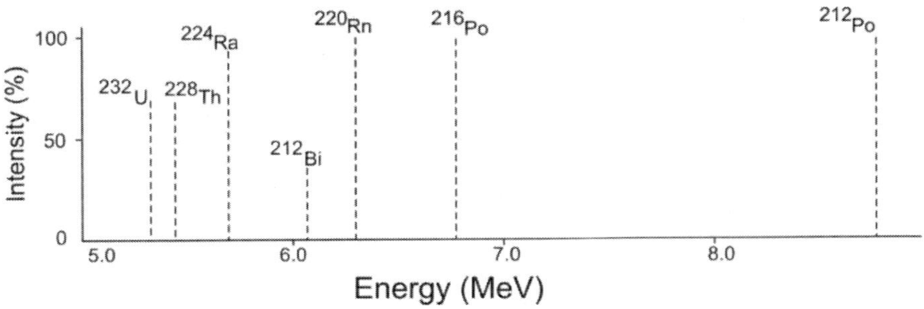

Figure 4. Alpha particles generated in the ^{232}U decay series. Data reported by [7].

Thus, the major steps involved in the extraction of the dissolved Th isotopes are:

- Qualitative filtration of the sampled volume for separating the suspended major particles and organic matter.
- Filtering in 0.45 μm Millipore membrane. The "soluble" thorium remains in the liquid phase, not in the filter. The filtration can be realized under vacuum conditions on putting a 47 mm diameter filter over a glass/porous support coupled to a 300 mL funnel by an anodized aluminum clamp.
- Sample acidification with 100 mL of hydrochloric acid for avoiding the iron precipitation and thorium adsorption in the reservoir walls.
- Addition of 200 μL of the ^{232}U-^{228}Th spike in radioactive equilibrium.
- Degassing of dissolved CO_2 for avoiding the formation of soluble complexes that could stop the uranium co-precipitation.
- Addition of ferric chloride (1 g) dissolved in 8M hydrochloric acid.
- Addition of ammonium hydroxide (200 mL) in order to raise the solution pH up to 13. Under these conditions, ferric hydroxide precipitates but not some aluminum complexes that are undesirable in the uranium deposition phase. The ferric hydroxide rapidly flocculates from a supersaturated solution, carrying uranium and thorium in the flocculate.
- Decantation of the solution during several hours.
- Filtering of the solution in a 124 mm diameter filter paper in order to recover the solid phase.
- Dissolution of the recovered phase with 8M hydrochloric acid.
- Iron extraction of the solution by isopropyl ether. The organic solvent and iron transfer to the ether phase after vigorous agitation. The uranium and thorium remain in the inorganic phase containing 8M hydrochloric acid.
- Insertion of the solution containing uranium and thorium in a column (1 m long and 12 mm diameter) containing an anions exchange resin of average porosity and total exchange capacity corresponding to 1.4 mEq/mL. The resin is strongly basic, is in the OH^- ionic form, and exhibits the radical R_4N^+ as an active group. The column receives 500 mL of solution that flows at a speed of 20 drops per minute. The resin is acidified with 8M hydrochloric acid (50 mL) for retaining the uranium/iron residual and releasing the thorium and earth alkaline elements. The solution containing thorium in 8M HCl, after passing

through the anion resin column, is subjected to the chemical treatment described below.

- Evaporation to dryness.
- Dissolution of the dry residue in 4 mL of distilled water.
- Addition of nitric acid (4 mL).
- Addition of distilled water up to a volume of 50 mL.
- Addition of 6-8 mL of ammonium hydroxide until a solution pH of 8.
- Heating of the solution up to the boiling point. Maintenance of this state during 2 minutes and rest for another 2 hours. Under these conditions, occur the Al and Th precipitation (like hydroxides) from the solution that passed through the resin.
- Filtering of the solution in a 124 mm diameter filter paper in order to recover the solid phase.
- Washing of the phase retained in the filter paper with 100 mL of a heated 3M sodium hydroxide solution, in order to remove aluminum.
- Washing of the residual phase retained in the filter paper with 100 mL of hot distilled water. The precipitate containing thorium remains in the filter paper.
- Dissolution of the residue containing thorium in 100 mL of heated 4M hydrochloric acid.
- Elution of the solution in a cation exchange resin. The resin is in the H^+ ionic form and its grain size varies between 200 and 400 mesh. The resin is previously acidified with 4M hydrochloric acid for separating the residual aluminum.
- Elution of the thorium retained in the resin by 125 mL of 0.75M oxalic acid.
- Addition of 125 mL of a 1:1 perchloric acid and oxalic acid mixture to the solution containing thorium. Thus, it occurs the oxalic acid decomposition according to the reactions:

$$3H_2C_2O_4 + 2HNO_3 \rightarrow 4H_2O + 2NO + 6CO_2$$
$$H_2C_20_4 + HClO_4 \rightarrow HClO_3 + H_2O + 2CO_2$$

- Dissolution of the dry residue in 2 mL of 0.1M nitric acid adjusted to pH between 3 and 4.
- Transfer of the solution to a 20 mL separation funnel.
- Addition of 2 mL of 0.4M thenoyl-trifluoroacetone diluted in benzene into the separation funnel. After agitation and mixing of the phases,

thorium transfers to the organic one, whereas the inorganic is discarded.

- Deposition of the organic phase containing thorium in stainless steel disc.
- Insertion of the disc in a muffle furnace at 500°C during about 30 minutes. The organic matter in the disc is eliminated, forming a thin and adherent thorium oxide layer, very appropriate to alpha spectrometry.

Figure 5. Simplified sketch map showing the location of the sampling points for groundwater.

General Features of the Study Area

The collection sites for groundwater were located at Poços de Caldas city, Minas Gerais State, and Águas da Prata city, São Paulo State, which are important spas in southeastern Brazil (Figure 5). In these areas, there is outcropping of volcanic rocks associated to the caldera subsidence of the Poços de Caldas alkaline massif. The complex is circularly shaped (Figure 6), with a mean diameter of 33 km. The total surface area is about 800 km^2, the altitude varies between 1300 and 1600 m, and the topography is characterized by valleys, mountains, and gentle grass-covered hills. The plateau is a ring structure of Mesozoic age comprising a suite of alkaline volcanic and plutonic rocks, mainly phonolites and nepheline syenites. The evolutionary history starts with major early volcanism involving ankaratrites (biotite-bearing nephelinite), phonolite lavas, and volcano-clastics, followed by caldera subsidence and nepheline syenite intrusions forming minor ring dykes and circular structures and, finally, the intrusion of eudialite-bearing nepheline syenites [10].

Water samples were collected at Águas da Prata city for analysis of the thorium isotopes in the following springs: Villela, Vitória, Platina, Prata Antiga, Prata Nova, Prata Radioativa, and Paiol. Villela spring discharges through vertical and horizontal open fractures into a well silicified and lightly folded sandstone [11, 12]. This spring is very much known in Brazil due to its high radioactivity related to dissolved ^{222}Rn [13]. Vitória spring discharges through fissures in diabase by ascendant water circulation [11]. Platina spring discharges through vertical fractures in outcropping phonolites by ascendant water circulation [11]. Prata Antiga spring was the first identified in the city and the water discharges through fissures and fractures in diabase [11]. Prata Nova spring corresponds in fact to some wells about 16 m deep which cut diabase and phonolites [11]. The water in Prata Radioativa spring discharges through fractures in a silicified and recrystallized sandstone [11]. The discharge of Paiol spring occurs through alkaline rocks like volcanic tuffs, phonolites and eudialite-bearing nepheline syenites [11].

Morro do Ferro is one of the highest points on the Poços de Caldas caldera, rising some 140 m above the base (Figure 7). Several boreholes were drilled to investigate the hydrological and chemical character of the subsurface flow on the aquifer system developed in the weathered profile of the Morro do Ferro area. Nine were drilled by IPT (Institute of Technological Researches) in 1982 to sample the upper part of the saturated zone, from which three in the ore body zone (SR-5, SR-7 and SR-8), four around the ore body zone (SR-2,

SR-4, SR-6 and SR-9) and two distant from the ore body zone (SR-1 and SR-3). Four other boreholes were drilled along the southeastern slope of Morro do Ferro to study the lower part of the saturated zone [14]: MF 10, situated in the ore body zone; MF 11, immediately adjacent to MF 10, to sample groundwater in the vicinity of the water-table for comparison with the deeper water sampled by MF 10; MF 12, at the bottom of the southeastern slope near the south stream, and MF 13, between boreholes MF 10 and MF 12. The water table is a subdued replica of the topography, with recharge on the higher elevations from precipitation and discharge locally to streams (North and South stream) (Figure 8) [15]. At the top of the hill, the water table is at least 80 m below the surface and in the valley bottoms is at or near the surface, occurring as seepages or discrete springs. Ground and surface waters from Morro do Ferro area were analyzed for dissolved thorium isotopes.

Results and Discussion

Spring Water Samples of Águas Da Prata City

Table 3 reports the results of the dissolved thorium content which are low as expected from the occurrence of its element in minerals of uneasy dissolution [1]. The highest value corresponded to 1.7 μg/L in Prata-Antiga spring, whereas the lowest (0.003 μg/L) was found in Villela spring that is well-known due to its enhanced levels in ^{222}Rn activity concentration [12]. Because ^{222}Rn is an intermediate nuclide in the ^{238}U-decay series, such finding suggests that are different the mechanisms of radionuclides dissolution in the ^{238}U and ^{232}Th-decay series.

In general, low dissolved thorium content has been reported in literature, for instance, 0.64 ± 0.20 μg/1000L and 4.5 ± 0.8 μg/1000L in ocean waters [19]. However, some authors reported values without considering the samples filtration like [20] who determined values between 0.02 and 0.24 μg/1000L in unfiltered samples. Such procedure may cause discrepant results that must be considered with caution due to contamination of the liquid phase by suspended thorium.

All waters analyzed in Águas da Prata area exhibited isotopic fractionation between ^{228}Th and ^{232}Th (Table 3). In some cases, high activity ratios values such as above 10 were found, as well also reported in other situations, for instance, ocean waters [21]. These ratios have been interpreted taking into account the possible preferential mobilization of ^{228}Ra relative to ^{228}Th that causes a ^{228}Th enrichment in the soluble and suspended solids fractions.

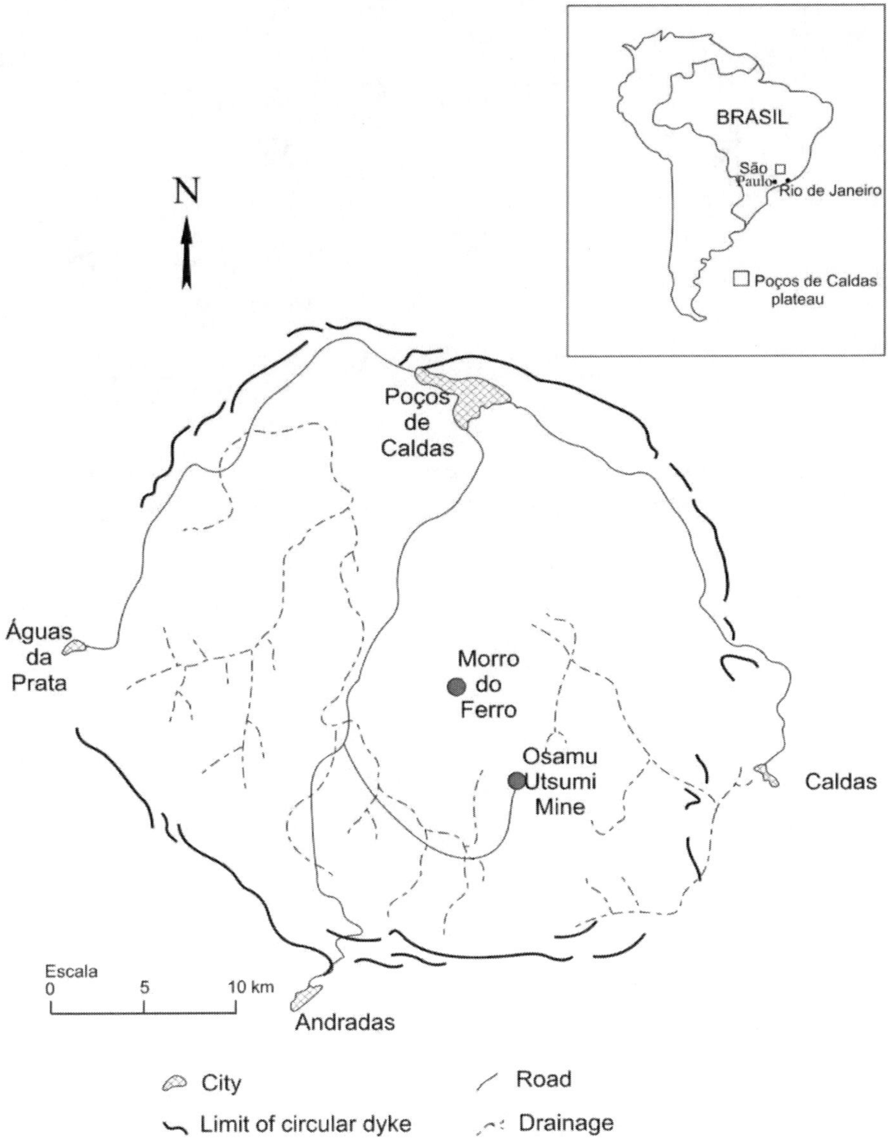

Figure 6. Location of Águas da Prata and Poços de Caldas cities at the Poços de Caldas alkaline massif.

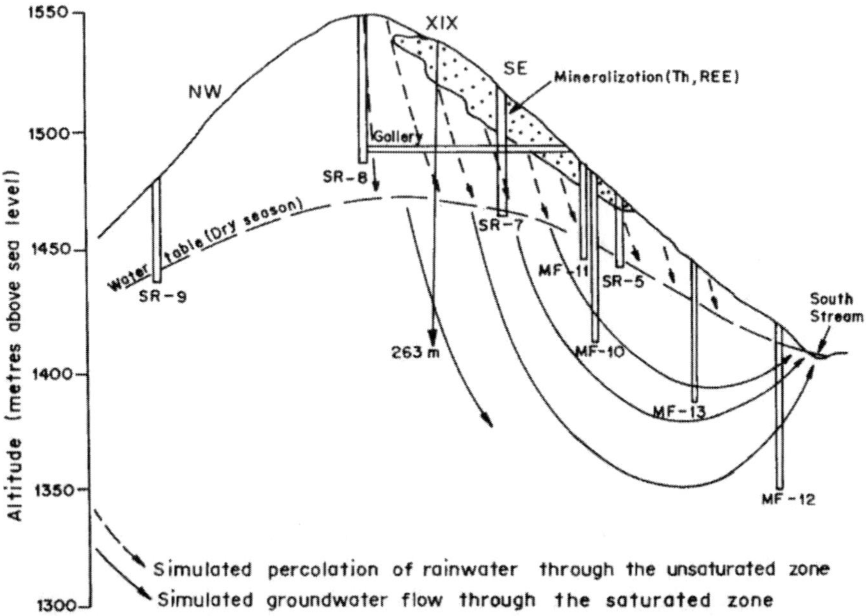

Figure 7. Profile across Morro do Ferro showing the mineralization of Th and REEs and the position of the water table during the dry season. Adapted from [16,17].

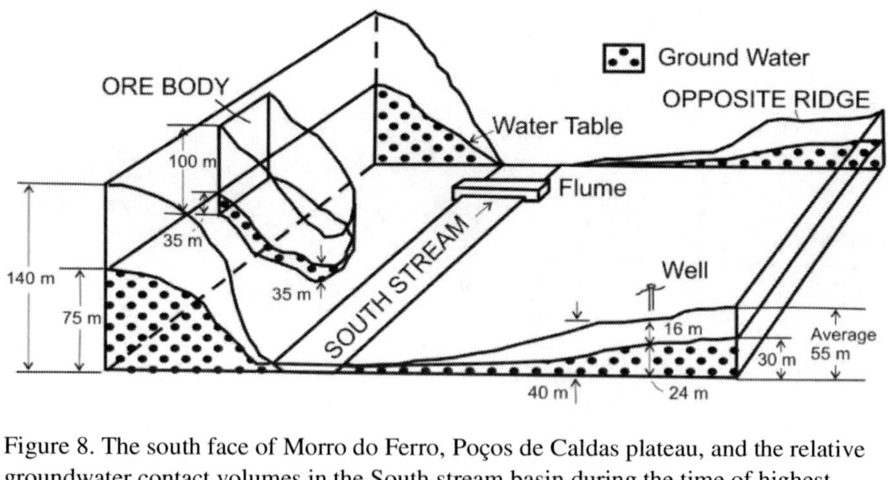

Figure 8. The south face of Morro do Ferro, Poços de Caldas plateau, and the relative groundwater contact volumes in the South stream basin during the time of highest water table level in the 1981/1982 rainy season. According to [18].

Table 3. Dissolved thorium content and $^{228}Th/^{232}Th$ activity ratio (AR) in groundwater occurring at Águas da Prata city, São Paulo State, Brazil. The analytical uncertainty corresponds to ± 10% at 1σ standard deviation. According to [22]

Sample name	Sampling date	^{232}Th (μg/L)	^{230}Th (dpm[1])	$^{228}Th/^{232}Th$ A.R.
Platina	21 Jun 1995	0.11	0.70	1.77
Paiol	21 Jun 1995	0.61	2.10	2.79
Near Paiol	13 Apr 1994	0.24	2.39	13.00
Prata Radioativa	13 Apr 1994	0.11	0.85	2.74
Prata Nova	21 Jun 1995	0.65	8.93	8.31
Prata Antiga	13 Apr 1994	1.72	23.45	5.85
Villela	21 Jun 1995	0.003	22.49	34.47

[1] disintegrations per minute.

The ^{230}Th activity data reported in Table 3 were obtained by the isotope dilution technique, where the highest value in Villela spring is possibly due to the presence of uranium in the filling material of cracks in the silicified sandstone that are able to store water [21]. Statistical tests of correlation were performed among dissolved thorium and anions HPO_4^{2-}, NO_3^-, SO_4^{2-}, Cl^- and F^- commonly verified in the analyzed waters as they may form inorganic complexes with Th in some circumstances [1]. However, none significant correlation was found as reported elsewhere.

Table 4 indicates that thorium in suspended matter was found in three groundwater samples. The results show an accentuated thorium mobilization in this phase, which may be adsorbed in organic or inorganic materials as pointed out by [1]. Thus, the average dissolved thorium content was 0.49 μg/L, whereas the mean thorium concentration in suspended matter was 1308 μg/g. These values allow estimate a factor of $\sim 3\times10^6$ for the preferential thorium mobilization in the particulate matter relative to the liquid phase.

Ground and Surface Water Samples of Morro Do Ferro Area

The $^{228}Th/^{232}Th$ activity ratios were calculated from the corresponding peaks in the α-spectra of the Th-bearing planchets. The $^{228}Th/^{232}Th$ activity ratio (AR) was calculated by the equation $^{228}Th/^{232}Th\ AR = N_{228}/N_{232T}$, where N_{228} and N_{232T} are the net counts in the ^{228}Th and ^{232}Th peaks, respectively.

The spike ^{228}Th was added to all samples collected in December 1982 and March 1983. Table 5 reports the results obtained, which indicate no chemical recovery of thorium for the groundwater sample of bore SR-3 collected in December 1982. Table 6 reports the results obtained for samples collected in September 1983. In this case, the ^{228}Th/^{232}Th activity ratio for spiked and non-spiked water samples of bore SR-3, bore SR-4 and flume are equivalent, within experimental errors, indicating no possibility of identifying the signal corresponding to the ^{228}Th spike added.

Figure 9 illustrates the alpha spectrum obtained for the groundwater sample of bore SR-1 collected in December 1982 (volume = 19 L; counting time = 63 hours). Such sample was not filtered in 0.45 μm Millipore membrane and, therefore, the thorium isotopes include the contribution of the soluble and particulate phases. The lower ^{228}Th/^{232}Th activity ratio for this sample in relation to others (Table 5) may reflect a more pronounced ^{232}Th transport in the particulate phase. The results of the analyzes performed by [18] for many ground and surface waters collected at Morro do Ferro area indicated that the particulate phase contains more than 90% of the transported thorium, indicating the importance of this phase in its mobilization.

The determination of the ^{232}Th concentration in waters from Morro do Ferro area was performed on analyzing two aliquots of each sample. The ^{232}U-^{228}Th spike was added to one aliquot and both of them were subjected to the same radiochemical procedure for Th extraction and deposition. The alpha spectrum for the non-spiked sample yielded a ratio $(N_{228}/N_{232T})_{nat}$. The net counts N_{228spk} due to ^{228}Th spike in the ^{228}Th peak can be obtained by the formula:

$$N_{228spk} = (N_{228}) - N_{232T} \times (N_{228}/N_{232T})_{nat}$$

where: N_{228} are the net counts in the ^{228}Th peak (spike + sample).

Table 4. Thorium concentration and ^{228}Th/^{232}Th activity ratio in suspended matter from groundwater occurring at Águas da Prata city, São Paulo State, Brazil. The analytical uncertainty corresponds to ± 10% at 1σ standard deviation. According to [22]

Sample name	Sampling date	^{232}Th (μg/g)	^{228}Th/^{232}Th A.R.
Near Platina	13 Apr 1994	3445	2.58
Vitória	13 Apr 1994	183	11.08
Villela	13 Apr 1994	297	6.45

Table 5. Alpha readings and respective $^{228}Th/^{232}Th$ activity ratio for ground and surface water samples collected at Morro do Ferro area, Poços de Caldas plateau

Water sample	Sampling date	Vol. (L)	Counting Time (hours)	N_{228} (counts)	N_{232T} (counts)	$^{228}Th/^{232}Th$ AR
Bore SR-1 (non-filtered)	12/22/1982	19	63.07	2339.86	231.16	10.12
Bore SR-2	12/22/1982	19	5.17	1369.89	38.84	35.27
Bore SR-3 (non-filtered)	12/22/1982	19	126.12	not detected	not detected	not calculated
Bore SR-4	12/22/1982	19	7.42	1547.99	41.67	37.15
Bore SR-5	12/22/1982	18	0.25	14998.78	19.99	750.31
Bore SR-6	12/22/1982	18	138.76	694.68	not detected	not calculated
Bore SR-9	12/22/1982	18	21.86	1438.62	not detected	not calculated
Bore SR-1	03/09/1983	19	121.46	4181.62	129.53	32.28
Bore SR-3	03/09/1983	19	19.60	2456.41	67.41	36.44
Flume (South stream)	03/09/1983	19	15.97	1712.30	62.52	27.39

Table 6. Alpha readings and respective $^{228}Th/^{232}Th$ activity ratio for ground and surface water samples collected at Morro do Ferro area, Poços de Caldas plateau

Water sample	Sampling Date	Vol. (L)	Counting Time (hours)	N_{228} (counts)	N_{232} (counts)	$^{228}Th/^{232}Th$
Bore SR-1 (non-spiked)	09/27/1983	17.6	20.24	9379.18	38.14	245.91
Bore SR-1 (spiked)	09/27/1983	17.6	1.11	14566.13	22.72	641.11
Bore SR-2 (non-spiked)	09/28/1983	16.8	17.28	12396.33	125.28	98.95
Bore SR-2 (spiked)	09/28/1983	19.4	15.39	11364.54	64.54	176.09
Bore SR-3 (non-spiked)	09/27/1983	17.6	24.15	1018.80	11.28	90.32

Table 6. (Continued)

Water sample	Sampling Date	Vol. (L)	Counting Time (hours)	N_{228} (counts)	N_{232} (counts)	$^{228}Th/^{232}Th$
Bore SR-3 (spiked)	09/27/1983	19.4	15.26	4054.64	34.31	118.18
Bore SR-4 (non-spiked)	09/27/1983	16.8	30.12	6984.86	38.00	183.81
Bore SR-4 (spiked)	09/27/1983	17.6	14.56	3579.20	21.56	166.01
Bore SR-5 (non-spiked)	09/28/1983	17.6	1.20	33671.48	58.07	579.84
Bore SR-5 (spiked)	09/28/1983	17.6	0.59	12867.68	50.57	254.45
Bore SR-6 (non-spiked)	09/28/1983	17.6	17.28	8301.56	40.42	205.38
Bore SR-6 (spiked)	09/28/1983	17.6	4.30	16581.93 1429.10	53.65	309.08
Bore SR-9 (non-spiked)	09/27/1983	16.3	15.07	19630.05	31.09	45.87
Bore SR-9 (spiked)	09/27/1983	17.6	13.96	9871.75	68.83	285.20
Flume (non-spiked)	09/28/1983	15.2	25.89	2710.75	76.41	129.19
Flume (spiked)	09/28/1983	20.0	13.64		24.65	109.97

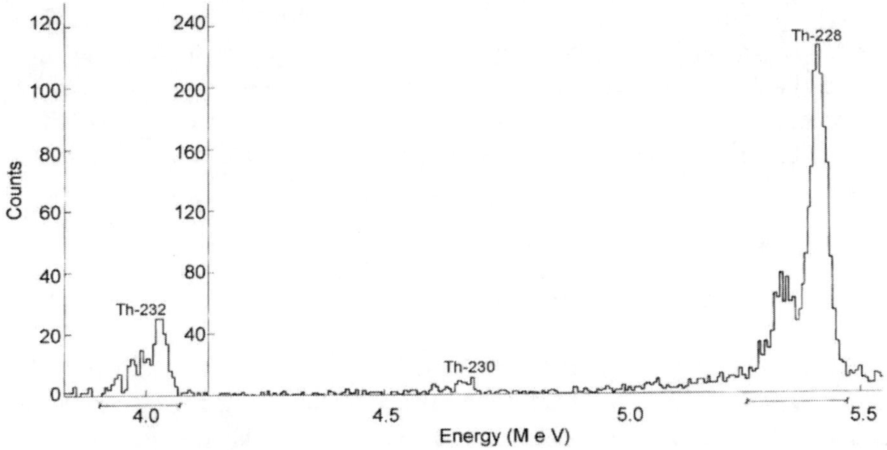

Figure 9. The alpha spectrum obtained for thorium extracted from groundwater sample of bore SR-1 drilled at Morro do Ferro, Poços de Caldas plateau. Sampling date = December 1982.

Since one microgram of ^{232}Th corresponds to an activity of ~ 0.246 dpm, the thorium concentration (C_{232}, in μg/L or ppb) may be obtained from the ^{232}Th activity, the sample volume (V, in liters) utilized for extracting thorium and the isotope dilution principle. Thus, it is possible to write:

$$C_{232} = Ao/0.246V\ (N_{232T}/N_{228spk})$$

where: V = volume of the spiked aliquot; Ao = activity of the ^{228}Th spike added to each sample (6.58 ± 0.29 dpm).

The ^{228}Th/^{232}Th AR for spiked and non-spiked water samples of bore SR-3, bore SR-4 and flume in Table 6 are equivalent, within experimental errors, indicating no possibility of identifying the signal corresponding to the ^{228}Th spike added. Therefore, it is not possible determine N_{228spk} for these samples and, consequently, the concentration of dissolved ^{232}Th. The ^{228}Th/^{232}Th AR in ground and surface waters ranged from 27 to 750 (Tables 5 and 6). These values greatly exceed unity, characterizing a high isotopic fractionation between these nuclides.

^{228}Th is a ^{228}Ra-daughter that is a β^--emitting nuclide whose half-life is 6.7 years. ^{228}Ra exhibits higher solubility than that of its parent, ^{232}Th. Because of its short half-life (1.9 years), the generation of ^{228}Th in detectable amounts occurs in the initial months after the ^{228}Ra-production in the solid phase and preferential transport to the water. This causes an “apparent” ^{228}Th-enrichment in the filtered phase that is not enough to impoverish it in the solid one. The same considerations were realized by [23] for the $^{228}Th/^{232}Th$ activity ratios in filtered aliquots from surface water samples collected at Morro do Ferro. They also confirm the findings of [24] who reported that radium is strongly leached by rainwater percolating the deposit so that the ^{228}Th produced by ^{228}Ra-decay remains in the solid phase.

CONCLUSION

The $^{228}Th/^{232}Th$ AR in waters collected in September 1983 at Morro do Ferro area, Poços de Caldas plateau, are generally higher than those found for samples collected in December 1982 and March 1983. The samples collected in September 1983 were not related to a dry period, as the rainy period started in that month, following the drier months of July and August 1983. A possible explanation for the higher values of the $^{228}Th/^{232}Th$ activity ratios could be related to the fact that the ^{228}Ra produced in the sediments and not leached during the dry season would be transported by the waters infiltrating during the initial rainfall period. Other studies held in the site verified the rainfall effect in the concentration of dissolved thorium in groundwater samples collected from bore SR-2 during February 1982. Data on the rainfall, dissolved Th content, sampling date and water table depth indicated the precipitation occurrence at the following periods: from 2/02/1982 to 2/08/1982 and from 2/21/1982 to 2/27/1982. The climate was dry between those epochs, the main short duration precipitation occurred in 2/02/1982 and 2/21/1982, whereas the others events were of low/moderate intensity rainfall that supplied less than 10 mm in one hour. The data indicated that the high rainfall occurring in 2/02/1982 did not increase the dissolved Th content. On the other hand, low intensity rainfall like occurring in 2/03/1982, 2/04/1982 and 2/05/1982 appeared to enhance the Th concentration in 2/06/1982, like also verified for the rainfall in 2/22/1982, 2/23/1982 and 2/24/1982 that were possibly related to the Th content increase in 2/25/1982. The high Th content verified in 2/25/1982 was attributed to dry conditions occurring in the soil cover, as well to low intensity rainfall able to produce infiltration water. Under these conditions, the percolation water would transport not only ^{232}Th but also complex agents that would help to maintain the thorium into solution. Natural fulvic and humic acids in groundwater occurring at the ore body zone in Morro do Ferro could be important agents responsible for Th complexation, thus, justifying its presence in the liquid phase as pointed out by other investigators. The values obtained for dissolved thorium in Morro do Ferro area allowed estimate a factor of ~5×10^6 for the preferential thorium mobilization in the particulate matter relative to the liquid phase that is almost twice of that estimated for the spring waters of Águas da Prata city. Despite both sites are located in the same alkaline massif, the different factors were probably found because the groundwater at Morro do Ferro area transports a more accentuated suspended load that implies on an enhanced thorium mobilization in the suspended solids phase relative to the liquid one.

REFERENCES

[1] Langmuir, D; Herman, JS. *Geochim Cosmochim Acta*, 1980, 44, 1753-1766.

[2] Ivanovich, M; Harmon, RS. Uranium Series Disequilibrium: *Applications to Environmental Problems*; Clarendon Press: Oxford, 1992.

[3] WHO (World Health Organization) Guidelines for drinking-water quality; 4th ed; WHO Press: Geneva, 2011.

[4] Ku, TL. *An Rev Earth Plan Sci*, 1976, 4, 347-379.

[5] Andrews, JN. Laboratory procedures; Bath University: *Bath*, 1984, 28

[6] Adams, JAS; Gasparini, P. Gamma ray spectrometry of rocks; Elsevier: Amsterdam, 1970, 279 pp.

[7] Chu, SYF; Ekström, LP; Firestone, RB. (1999). The Lund/LBNL Nuclear Data Search. http://nucleardata.nuclear.lu.se/nucleardata/toi/ index.asp

[8] Bonotto, DM. Aplicações hidrogeoquímicas dos isótopos naturais do U (4n+2) e Th (4n) no Morro do Ferro, Poços de Caldas (MG); PhD Thesis; Instituto Astronômico e Geofísico – USP: São Paulo, SP, 1986.

[9] Mertzig, W. Estudo da técnica de eletrodeposição na preparação de amostras para determinação de ^{234}U por espectrometria alfa; M. Sc. Dissertation; Instituto de Pesquisas Energéticas e Nucleares: São Paulo, SP, 1979.

[10] Ellert, R. Bol FFCL USP (Geol) 1959, 237 (18), 5-64.

[11] Szikszay, M. Hidrogeoquímica das fontes de Águas da Prata, Estado de São Paulo; Post PhD Thesis; USP: São Paulo, SP, 1981.

[12] Szikszay, M; Sampa, MHO. *Boletim IG*, 1982, 13, 25-42.

[13] Longo, OW. *Rev Inst Geogr Geol*, 1967, 19, 27-48.

[14] Holmes, DC; Pitty, AE; Noy, DJJ. *Geochem Explor*, 1992, 45 (1-3), 215-247.

[15] Eisenbud, M; Cowan, GA; Duguid, JO; Krauskopf, K. Findings and recommendations concerning the thorium deposit o the Moro do Ferro; New York University Medical Center: New York, 1979, 21 pp.

[16] IPT (Instituto de Pesquisas Tecnológicas do Estado de São Paulo S/A) Estudos para a caracterização do comportamento do lençol freático na área do Morro do Ferro - Poços de Caldas (MG); DMGA: São Paulo, 1982, Rep. 16423 (22), 16 pp.

[17] Waber, N; Schorscher, HD; Peters, TJ. *Geochem Explor*, 1992, 45 (1-3), 53-112.

[18] Lei, W. Thorium mobilization in a terrestrial environment; PhD Thesis; New York University Medical Center: New York, NY, 1984.

[19] Moore, WS; Sackett, WM. *J Geophys Res*, 1964, 69, 5401-5405.

[20] Somayajulu, BLK; Goldberg, ED. *Earth Planet Sci Lett*, 1966, 1, 102-106.

[21] Moore, WS. *Earth Planet Sci Lett*, 1981, 53, 419-426.

[22] Tonetto, ÉM; Bonotto, DM. *Rev Brasileira Geociências*, 2002, 32, 343-350.

[23] Eisenbud, M; Lei, W; Ballad, R; Franca, EP; Miekeley, N; Cullen, T; Krauskopf, K. In Scientific Basis for Nuclear Waste Management V; Lutze, W.; Ed.; North-Holland: New York, NY, 1982, 735-744.

[24] Franca, EP; Campos, MJ; Carlos, M; Lobão, N; Pfeiffer, WC; Trindade, H. Radium, leaching, transport and retention at Morro do Ferro, Poços de Caldas, MG; CNEN: Rio de Janeiro, 1983, INIS-BR-86, 261-262.

In: Thorium
Editor: Renae Jamison
ISBN: 978-1-63321-309-8

Chapter 6

THORIUM OCCURRENCE: GEOLOGICAL AND GEOPHYSICAL IMPLICATIONS FOR WESTERN AUSTRALIA

M. F. Middleton, J. Brett and D. Flint
Department of Mines and Petroleum
Western Australia, Australia

ABSTRACT

This study reviews thorium occurrence in terms of geology and resources in Western Australia. Australia, it has been reported, has some of the highest thorium (Th) concentrations recorded in rocks in the World. High thorium concentrations have been recorded in rocks in Western Australia for over fifty years, and the association of thorium with heavy-mineral-sand deposits has been recognized during this time in the Perth and Carnarvon basins. Studies have also recognized high thorium levels in the granitoids (geological bodies with mineral composition similar to granite) of the Yilgarn and Pilbara cratons of Western Australia. Recent investigations have detected unusually high levels of thorium in granitic gneiss of the Leeuwin Complex in the southwest of Western Australia, and supported high thorium concentrations in the Northampton Complex north of the Perth Basin, which may contribute to high heat flow in these regions.

The study shows known thorium distribution in Western Australia from both geological, geochemical and geophysical information. It reviews, especially based on airborne geophysics, a number of the higher thorium-concentration locations in the state and their genesis. Further, a detailed discussion of the impact of thorium on heat generation in the upper crust in Western Australia suggests that this element plays a significant role in the terrestrial heat flow in various regions, and may have implications for future geothermal energy generation.

Thorium is present as a minor constituent in many minerals, but particularly monazite, and is widely scattered across Western Australia. The Department of Mines and Petroleum's database of mines, deposits and prospects in Western Australia (MINEDEX) records 110 occurrences where monazite is listed as a commodity (mineral) for a particular site. Most of these are related to heavy mineral sands along the Perth Basin, both north and south of Perth. The concentration of monazite within heavy mineral sands is highly variable, but can be up to about 3%. While there is no commercial thorium production currently in Western Australia, monazite concentrate was produced from 1963 to 1994 as a by-product from mining zircon and ilmenite. Monazite by-product is also possible from mining tantalite in pegmatite at, for example, the Wodgina and Greenbushes deposits.

The presence of thorium, together with uranium and potassium, is known to contribute to heat generation within the Earth. Thorium, together with uranium (U), distributed throughout the acidic (granitoid association) rocks in the middle and upper Earth's crust, may contribute to as much as 75% of the observed heat flow out of the Earth's interior. Th^{232} has been reported to have a heat production of approximately 0.07 μWm^{-3} versus U^{238} having approximately 0.25 μWm^{-3}. Generally in Western Australia, thorium is about two to four times more abundant than uranium, which makes its contribution to heat generation almost equivalent to, and sometimes greater than, the contribution of uranium. It is estimated that, when $Th/U > 3.7$, then thorium will have a greater contribution to heat generation than uranium. Interestingly, recent studies in the southwest of Western Australia have shown that at some localities thorium is very much more abundant than uranium (with the ratio Th/U in the range of 10 to 15).

In terms of specific concentrations in Western Australia, thorium concentrations in granitoids have been measured between 170 to over 250 ppm thorium from ground-calibrated-geophysical gamma-spectrometer surveys; corresponding uranium concentrations varied between 8 and 16 ppm. Consequently, these data appear to show that thorium can have greater importance than uranium in heat generation and heat flow in many parts of Western Australia. The comparison of surface thorium measurements with respect to recently acquired aero-radiometric and aero-magnetic data in Western Australia exhibit previously unrecognized correlations for many regions in Western Australia; some of these are reported in this review.

A detailed analysis of the growing thorium database indicates a strong inter-relationship between thorium concentration, heat generation and concomitant natural terrestrial heat flow in the Leeuwin, Gascoyne and to a lesser extent in parts of the Yilgarn and Pilbara regions of Western Australia.

INTRODUCTION

Western Australia comprises approximately 30% of the Australian continent (Figure 1). The relevance of the element thorium in geological processes in this state is often unrecognized. Further, thorium abundance in the Earth is poorly known, but is cited as being more abundant than tin and uranium, and about as abundant as lead (Wickleder et al., 2006). Its presence contributes significantly to many of the processes that formed, and continue to form, the heat flow regime in the formation of the Earth, and may also play a substantial role in society's future energy supply, both nuclear and geothermal.

A 2008 publication by Geoscience Australia (Mernagh and Miezitis, 2008) has proposed that Australia possesses the largest total identified thorium resources in the world (18%), followed by the United States (16%) and then Turkey (13.8%). This has been disputed by some sources, which places India and the United States as containing greater resources. Nevertheless, it is valuable to recognize that very little knowledge is extant about thorium distribution on the planet.

With regard to Australia, Mernagh and Miezitis (2008) suggest that 64% of thorium resources (principally, "heavy mineral and other deposits") reside in the state of Victoria, 20% in Western Australia, and the remaining distributed between the other states and territories in Australia (see insert in Figure 1). These appear to be local concentrations of thorium, because continent-wide mapping of surface thorium in Australia from aero-radiometric surveying indicates a substantial surface concentration in Western Australia (RMA, 2010).

The presence of a significant areal distribution of high surface levels of thorium concentration in Western Australia perhaps also suggests crustal tectonic influences on thorium distribution in the Earth's crust and possible geothermal energy impacts.

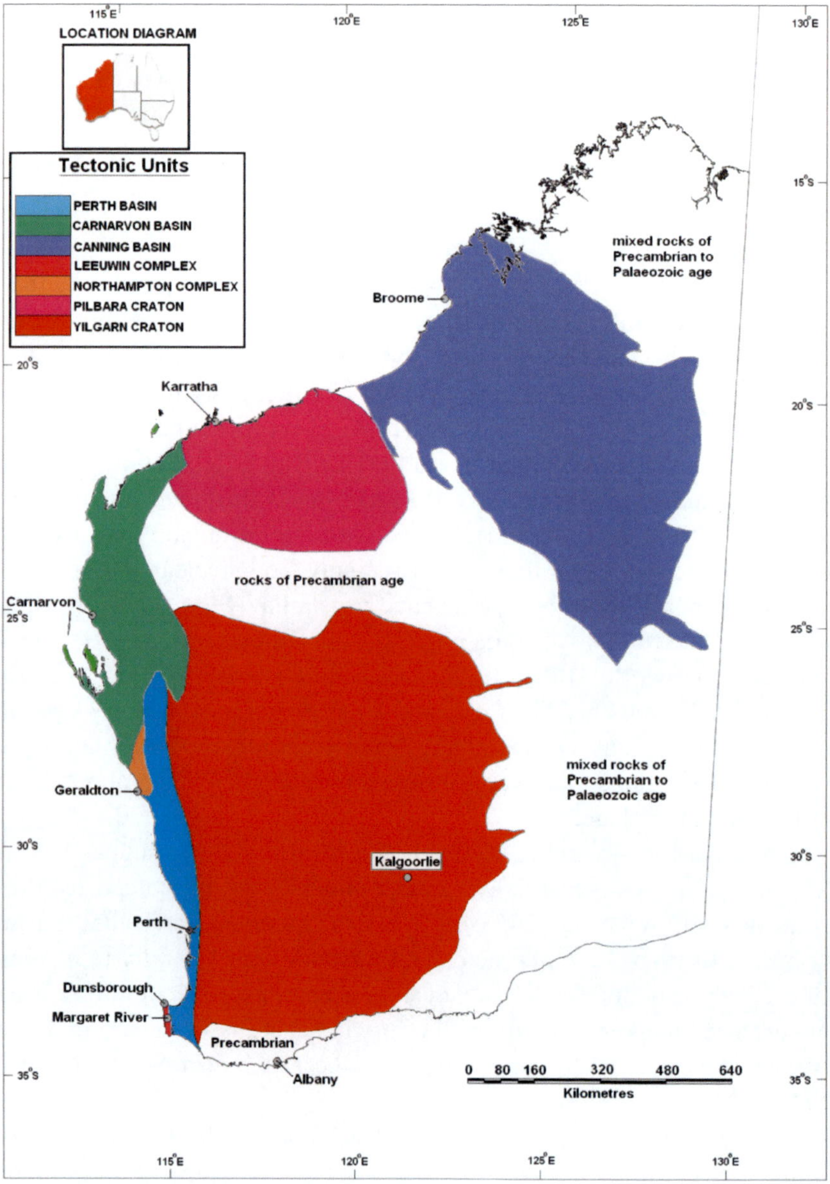

Figure 1. Map of Western Australia showing the major geological provinces and areas of special interest.

This paper reviews aspects of thorium distribution in Western Australia with respect to mining, thermal-sourced petroleum generation and geothermal energy. Thorium-rich minerals have been mined over the past 20 years. Radioactive-rich minerals have been recognized and mapped by petroleum exploration companies since the 1960s, and these are commonly associated with monzonite-rich sand deposits in the sedimentary basins on the western margin of Western Australia. Recent studies by Middleton (1913) and Middleton and Stevens (2013) have shown that thorium and uranium content of basement granitoids (geological bodies with mineral composition similar to granite) can strongly influence heat flow and subsurface temperature in the south west of Western Australia. This is an unexpected result, and appears to have implications for the understanding of the geological evolution of Western Australia and its potential geothermal resources.

Thorium was reported in Western Australia during the 1920s (Simpson, 1928) as a principal element in various minerals in the Wodgina mineral region. It was assumed at the time that thorium, contained in the accessory minerals Mackintoshite, Thoroguminite, Pilbarite and Hydrothorite, at the levels found at the Wodgina sites was not of economic importance.

Detailed geochemical analyses of the various regions of Western Australia have been published since 2000 (e.g. supportive mapping for the Ajana 1:250,000 geological sheet by Sanders and McGuiness, 2000). Studies, such as these, often present a detailed map of thorium distribution in the mapped region. An extensive data base of geochemical information exists in a data base maintained by both the Geological Survey of Western Australia and Geoscience Australia (Budd et al., 2011).

The present review presents some thorium-rich mineral deposits, geophysical mapping of surface thorium, and geothermal implications for thorium distribution in Western Australia. A brief mention is made of the impact of thorium in the context of Western Australian tectonic genesis, although much of this is speculative at this stage. An important outcome of this review is that anomalous concentrations of thorium have been detected in various regions of Western Australia, especially in the south west Leeuwin Complex and the Northampton Complex, impacting on heat flow in the southern Perth Basin and northern Perth Basin, respectively. The occurrence of high levels of thorium in these regions is currently unexplained.

THORIUM-RICH MINERAL DEPOSITS IN WESTERN AUSTRALIA

Boyle (1982) reviewed thorium-rich minerals in Western Australia, and a recent update for all of Australia has been published by Mernagh and Miezitis (2008). Common minerals containing thorium are monazite, thorianite, thorite and thorogummite. Monazite is the principal ore of thorium, and occurs in many sedimentary, igneous and metamorphic rocks in Western Australia. Monazite is the main thorium-bearing mineral being mined in Western Australia, although it is not being mined for its thorium content. These deposits largely occur in heavy-mineral-sand deposits in the south west of Western Australia (Baxter, 1977). The location of these deposits are clearly defined by aero-radiometric mapping (Brett, this volume; Brett, 2013), and a display of the aero-radiometric data for the Southwest of Western Australia is shown in Figure 2.

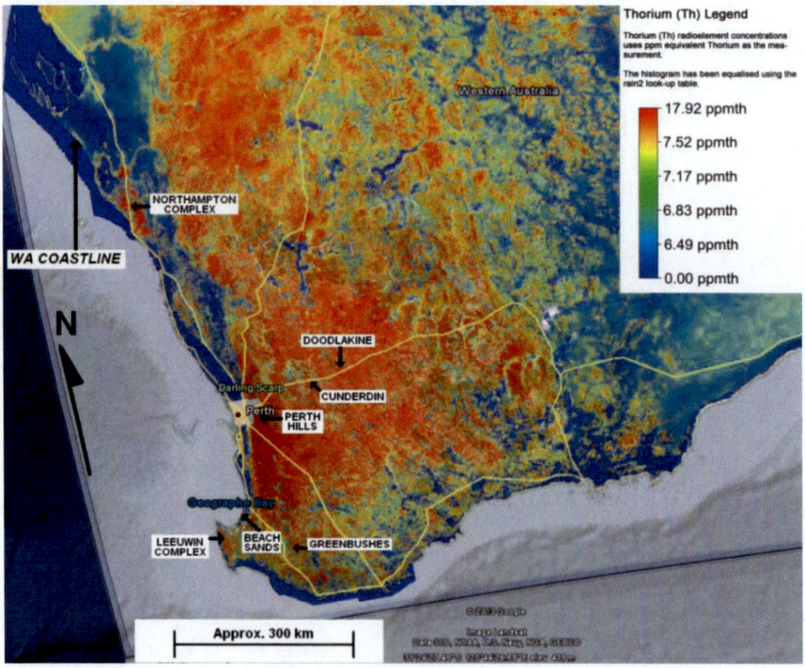

Figure 2. Thorium merge map of western Australia, showing calibrated surface thorium concentration , based on aero-radiometrics (Brett, 2013). Locations of specific interest in this review are also shown. It is acknowledged that the thorium merge map is superimposed on a 2013 Google Earth map.

The Department of Mines and Petroleum's database of mines, deposits and prospects in Western Australia (MINEDEX) records 110 occurrences where monazite is listed as a commodity (mineral) for a particular site. Most of these are related to heavy mineral sands along the Perth Basin, both north and south of Perth.

Some specific thorium deposits are now reviewed in summary: (a) Heavy Mineral Sands in the South West of Western Australia, (b) Pilbara Region (Wodgina Deposit and nearby localities), (c) Greensbushes, and (d) Northampton Complex.

Wodgina Deposit and Other Pilbara Localities

The Pilbara region of Western Australia is known to be mineral rich (see Figure 1 for location). Commodities such as iron, copper, lead, zinc, gold and tantalum are found and mined in this geological province. These minerals often have thorium associations, and have been specifically reported in the Wodgina region near Marble Bar (Simpson, 1928). The Wodgina deposit is being mined for tantalum, but also has well known thorium-mineral associations. This deposit has a very close association with the granitoid rock-unit called the Numbana Monzogranite, which aero-radiometric data, and Geological Survey of Western Australia, indicates to possess anomalously high thorium and uranium concentrations.

The Wodgina 1:100,000 geological sheet (Geological Survey of Western Australia) shows the extent of outcrops of the Numbana Monzogranite in the region (Blewett and Champion, 2005; Hickman, 2012). The Wodgina tantalum deposit is located within this region and the characteristics of the deposit have been reported by Simpson (1928). The deposit also has associated thorium minerals. Simpson (1928) reported that the thorium minerals present were Mackintoshite, Thoroguminite, Pilbarite, Hydrothorite. These form a genetically related series of minerals of which the parent is the mineral provisionally called Mackintoshite.

The composition of ThO_2 in these minerals is reported (Simpson, 1928) in percent as:

Macintoshite:	24.72 %
Thorogummite:	24.46 %
Pilbarite:	31.34 %
Hydrothorite:	57.79 %

No current mining enterprises to extract thorium are occurring in the Pilbara region of Western Australia. However, some potential to extract thorium, commercially in the future, may exist in the Numbana Monzogranite and the greater Wodgina region of the Marble Bar 1:250 000 geological sheet (see below for location: Figure 11).

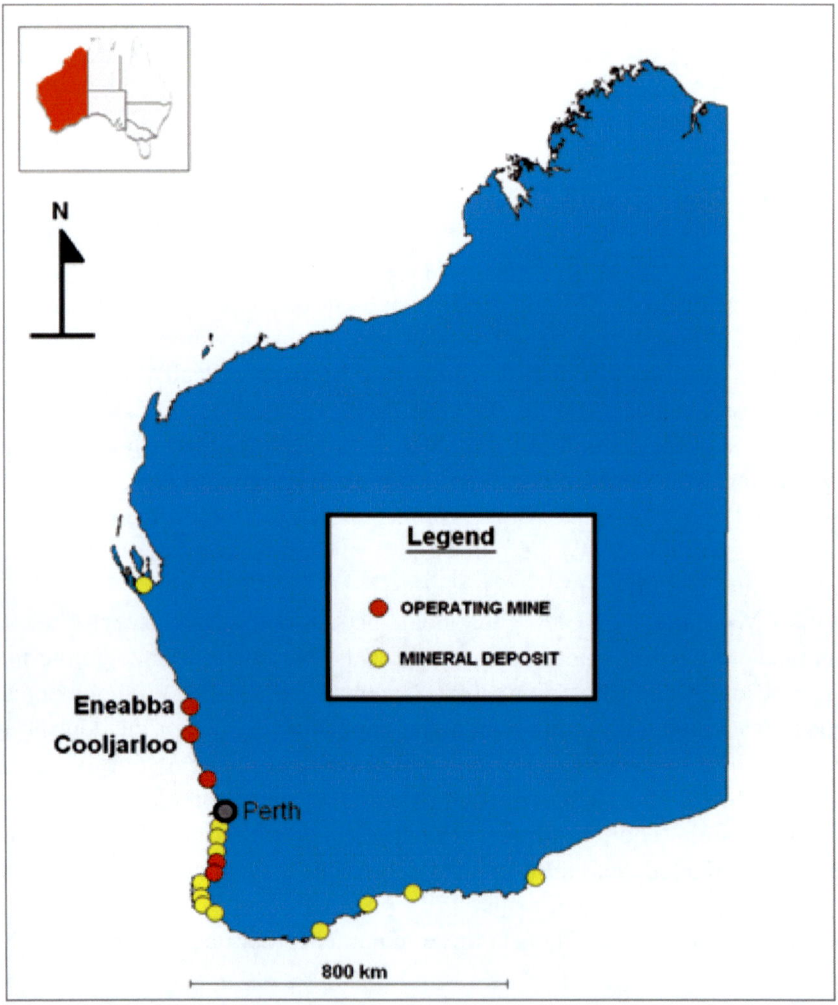

Figure 3. Map showing major mineral sand deposits and associated mines in the southern part of Western Australia.

Greensbushes and Southern Perth Basin Heavy-Mineral Sands

The Perth Basin contains rich-heavy-mineral sand-deposits (Figures 1 and 3). These deposits contain the mineral monzonite up to a concentration of about 1.1% (Mernagh and Miezitis, 2008). These authors have also indicated that there is an absence of thorium analyses from these deposits. They report that the maximum thorium composition constituted about 27%, but this being stated it should also be recognized that 95% of the grains analysed had less than 14.4% thorium. In conclusion, it is recognized that we know that thorium resides in the heavy-mineral-sand deposits in the Perth Basin, but little is known of the levels of concentration.

A maximum concentration of 138.9 ppm, from a mean of 18.1 ppm, derived from fine grained (< 0.05 mm grain size) stream samples measured from the Yilgarn Craton near the Greenbushes Tantalum deposit is reported. Interestingly, no thorium was detected in samples < 0.2 mm grain size. No explanation appears obvious for this at our present knowledge.

Northampton Complex

Little recent geological work has been carried on the Northampton Complex (GSWA, Memoir 3, 1990). The Northampton Complex (see Figure 1 for location) has been mapped on the Geraldton and Ajana 1:250,000 geological sheets (Hocking et al., 1982), and is part of the larger Pinjarra Orogen, which lies along the western margin of onshore Western Australia (Myers, 1990b).

On the Ajana 1:250,000 geological sheet, the Northamption Complex is dominated by rocks described as granulite and gneiss. These rocks have been metamorphosed under pressures in the range 600 to 900 MPa (6 – 9 kBar) and temperatures in the range of 600 to 800 °C (upper amphibolite to granulite facies metamorphic conditions: Sanders and McGuiness, 2000; Turner, 1968). Under these conditions, thorium and uranium mobility may be expected to be influenced, although little is known of this in this specific locality. No thorium analyses known to the authors have been reported in the literature for the Geraldton 1:250,000 geological sheet.

Sanders and McGuiness (2000) reviewed the geochemistry of the Ajana 1:250,000 geological sheet, and therein presented valuable data concerning the surface distribution of thorium in the regolith. For the purposes of this study, the regolith is defined as unconsolidated or indurated weathered rock, and includes residual and transported material that can cover and obscure underlying bedrock.

It is important to note that the regolith, as a geological unit, does not in general constitute freshly outcropping rock. However, elements may be concentrated or diluted, depending on the processes that formed the regolith. Thorium is often considered not to be mobile (Mernagh and Miezitis, 2008), and as such probably concentrates during regolith-forming processes. Interestingly, uranium is considered more mobile, and possible becomes less concentrated during the regolith-forming process.

The thorium concentrations obtained in the study of Sanders and McGuiness (2000) were measured by recognized laboratory methods. Their study has presented a systematic grid of surface thorium concentrations in outcropping rocks at a spacing of approximately five to ten kilometres, nominally one sample per 16 km^2. The map display indicates a range of thorium concentration between approximately 5 to > 39 ppm. Unfortunately, the upper limit of thorium concentration, although acknowledged to be greater than 39 ppm, is not reported in their review.

Summary of Mineral Exploration for Thorium in Western Australia

At present there is little economic incentive for thorium production. This will depend on a social and political desire for thorium-based nuclear energy into the future. Apart from the nuclear energy industry, thorium may provide a low-temperature geothermal energy source for special geological environments; these do occur in the south west of Western Australia, and are currently being investigated by the Department of Mines and Petroleum. More details of this potential are reviewed in subsequent sections.

Geophysical Mapping of Thorium in WA

History of Airborne Radiometric Mapping in WA

Geophysical mapping using radiometric data has been ongoing in Western Australia over the past half century. Radiometric maps have been published by Geoscience Australia over the past decade (RMA, 2010), and have shown surface thorium and uranium concentrations over most of Western Australia, based on a calibration devised by Geoscience Australia. An airborne geophysical survey, conducted in 2011, completed the uranium, thorium and potassium coverage of the state of Western Australia. Based on the pre-2011 digital maps, Middleton (2011) derived a surface heat generation (A_o) map, which is based on the equation:

$$A_o = 0.26\, C_U + 0.07\, C_{Th} + 0.10\, C_K, \qquad (1)$$

In this equation, heat generation (A_o) is in units of μWm^{-3}, and C_U is the concentration of uranium in ppm, C_{Th} is the concentration of thorium in ppm, and C_K is the concentration of potassium in percent. The equation shows the importance of uranium and thorium in terrestrial heat generation, and is discussed further below.

The Pre-2011 Radiometric Surveys

Airborne geophysical surveys have been frequently conducted previously to 2011in Western Australia. These have been a combination of both exploration company surveys and pre-competitive surveys by the Geological Survey of Western Australia (State) and Geoscience Australia (Federal). These surveys acquired spectral gamma-ray data, from which approximate surface thorium, uranium and potassium concentrations were derived. A summary of these data were produced in the products of Geoscience Australia (RMA, 2010).

The 2012 Thorium Map of WA

In 2011, a survey was conducted by the Geological Survey of Western Australia and Geoscience Australia to complete the airborne radiometric coverage of Western Australia. This survey did not cover metropolitan Perth, the capital city of Western Australia. Based on these data, a "thorium merge" map has been generated by Brett (2013), which illustrates areas of elevated thorium in Western Australia (Figure 4). Merge radiometric maps are a composite of various radiometric maps, both from industry and government, and are calibrated to an approximately common base level. Together with the thorium merge map, Brett (2013) has also generated uranium, potassium and ternary concentration merge maps for the Earth's surface in Western Australia.

Brett (2013) describes the process in generating this map. The Western Australian thorium map highlights a number of regions with anomalously high thorium.

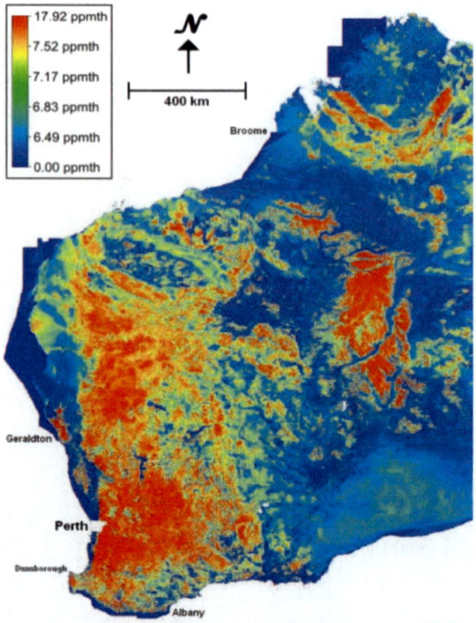

Figure 4. The thorium merge map for Western Australian (albeit the southern part) as can be best resolved through conventional software from of Brett (2013).

The thorium concentrations shown on the airborne-based thorium maps from (Figure 4) are calibrated to fall within expected thresholds determined by Geoscience Australia (RMA, 2010). It should be noted that in using equivalent uranium (eq U) and equivalent thorium (eq Th) to establish the thresholds in these maps, it is meant that that the amount of these two elements implied by Bi^{214} and Tl^{208} gamma radiation if equilibrium is assumed. However, equilibrium in the top few metres of the Earth is not a normal condition, especially for uranium, but for thorium it is not such an issue, because it is not as mobile as uranium. These threshold limits are between 0 and 19.5 ppm for thorium. However, it needs to be recognised that these values may not reflect surface values at any particular locality. This discrepancy is usually due to (1) the averaging effect of the collection radioactive emissions from a variety of ground radioactive sources by the airborne spectrometer detector, and (2) the integration of the flight-line traverse data into a pixel size of 80 m x 80 m. It is in recognition of this that surface "ground-truth" studies have been carried out by the Department of Mines and Petroleum of Western Australia to verify the aero-radiometric merge data (Middleton, 2012; Middleton, 2013). Five anomalous areas were identified for this study as having potential thorium concentration in the range 15 to 20 ppm, where 20 ppm is the threshold of the thorium merge-map calibration process, and are discussed below.

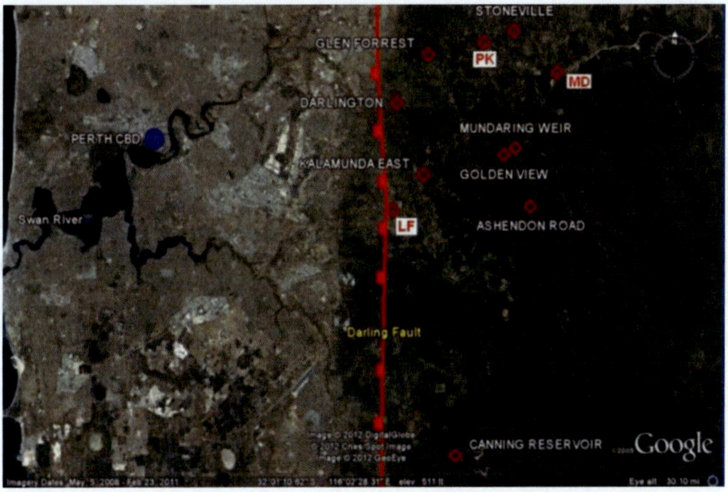

Figure 5. Map showing the location of measurements carried out by Middleton (2013) and additional measurements (initially reported in this paper) in the Perth Hills (Darling Range region), about 30 km west of the Perth CBD. New measurement localities are highlighted as PK (Parkerville), MD (Mundaring) and LF (Lesmurdie Falls).

Ground-Based Thorium Measurements

Thorium concentration can be made from laboratory or field measurements. Each method has its advantages and disadvantages. Laboratory methods are generally more accurate, but only represent the sample measured and can be relative more expensive. Field measurements using a spectral scintillometer (gamma-ray spectrometer) can be less accurate than laboratory determinations, but a greater area can be rapidly surveyed with a greater sample volume per measurement, and at less cost. The latter method was chosen for many locations reported in this study, because of time and economic factors.

Field measurements of abundance, or concentration, of naturally occurring radiogenic elements were made, initially in 2011, in order to estimate radiogenic heat generation of granitoids in the Darling Range region of the Yilgarn Craton. In an initial Darling Range study, thirteen sites (reported within Table 1; Figure 5) were reported by Middleton (2013). Further sites have been subsequently investigated, and are reported in this review. At each site, measurement of U, Th and K concentrations of outcropping granite rocks were carried out with a commercial RS-125 spectrometer, and complementary measurements of total disintegration counts were also made with a Geiger Müller counter. Both instruments sample the half-space of outcropping rock that they are placed upon. Thus, a flat surface of outcrop that has a minimum radius of 1 m is required for optimal measurement accuracy for the RS-125 (see Figure 6).

Table 1. Tabulation of mean, with standard deviation, values of uranium, thorium and potassium for various localities in the Yilgarn and Pilbara cratons in Western Australia. These data are for granite outcrop locations only

Location	Uranium (ppm) Mean, SD	Thorium (ppm) Mean, SD	Potassium (%) Mean, SD	No. of measure-ments	Heat Generation (μWm^{-3})
Stoneville	6.9, 0.5	32.6, 3.0	3.9, 0.1	3	4.5
Stoneville*	16, na	35, na	4.3, na	na	7.0
Ashendon Rd	7.3, 0.4	44.7, 1.8	3.8, 0.5	4	5.4
Golden View 1	11, 1.4	36.8, 3.0	4.1, 0.2	7	5.9
Golden View 2	9, 1.8	42.1, 4.2	4.5, 0.4	6	5.7
Golden View 3	6.7, 0.8	33.6, 1.4	4.6, 0.2	3	7.0
Darlington	6.1, 1.2	20.7, 1.5	2.5, 0.2	7	3.3
Canning Dam	7.3. 0.9	30.2. 1.1	2.6, 0.2	4	4.3
Kalamunda East	22.7, 1.0	55.7, 2.0	3.8, 0.2	6	10.2
Lesmurdie Falls (LF)	5.7, 0.8	54.3, 3.5	4.2, 0.2	6	6.1
Glen Forrest 1	17.0, 1.0	45.0, 3.4	4.0, 0.2	4	8.0
Glen Forrest 2	10.1, 1.7	42.6, 3.1	3.5, 0.1	3	6.0
Glen Forrest 3	15.8, 1.3	40.1, 3.3	3.7, 0.3	6	7.3
Glen Forrest 4	12.95, 1.24	39.2, 2.8	3.78, 0.4	6	6.5
Mundaring Weir	6.1, 0.9	34.4, 1.8	4.1, 0.2	8	4.4
Parkerville (PK)	9.4, 1.7	42.2, 3.7	3.8, 0.1	8	5.8
Doodlakine	10.4, 1.4	76.7,3.6	4.0, 0.3	12	8.5
Cunderdin	5.0, 2.4	65.2, 2.9	4.5, 0.1	5	6.3
Numbana (Pil)#	31.3, na	61.3, na	4.5, na	na	12.8
Coolegong (Pil)#	19.3, na	48.7, na	5.0, na	na	8.8

na: not available. * Data from Hot Dry Rocks Pty Ltd (2008).
#Data from Van Krandendonk (2011) for the Pilbara region.

Figure 6. Schematic diagram showing notional sampling volume for the RS-125 gamma-ray spectrometer. Essentially, a half sphere of about one metre radius is sampled for thorium, uranium and potassium isotopes in each measurement. The instrument is calibrated such that a cumulative sample time of 5 minutes (300 seconds) can provide a good match to laboratory-based thorium and uranium measurements.

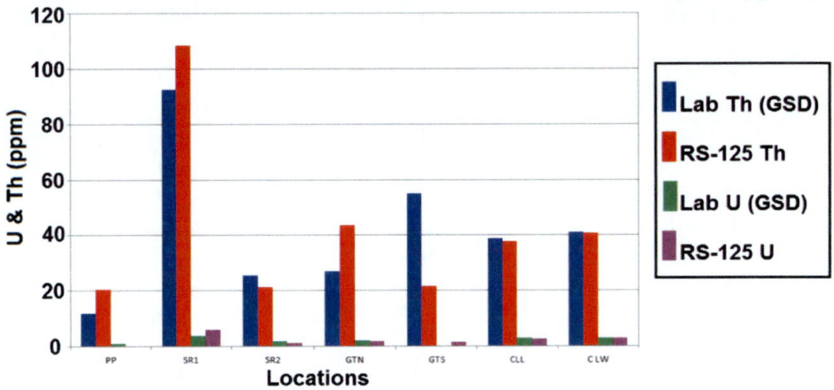

Figure 7. A plot of laboratory-based thorium and uranium measurements versus measurements from the RS-125 spectrometer. The comparison is for Leeuwin Complex measurements (Table 2). The codes on the plot refer to specific locations; pp = Point Picouet ; SR1 = Sugar Loaf Rock 1; SR2 = Sugar Loaf Rock 2; GTN = Grace Town North; GTS = Grace Town South; CLL = Cape Leeuwin Lighthouse; CLW = Cape Leeuwin Lighthouse west. In most cases, a good match is observed. These locations can be found on maps in Middleton and Stevens (2013).

Table 2. Mean U, Th and K concentrations, with the derived heat generation, measured at twelve localities in the Leeuwin Complex, and adjacent Vasse Shelf of the Perth Basin

Uranium (ppm)	Thorium (ppm)	Potassium (%)	Heat Generation (μWm^{-3})	Comment
6.0	108.4	7.4	9.9	Sugarloaf Rock (SR); Leeuwin Complex; previous GSWA site
2.1	37.9	5.7	3.8	Leeuwin Complex; previous GSWA site
2.8	37.8	5.6	3.9	Leeuwin Complex; previous GSWA site
3.3	46.8	6.2	4.8	Leeuwin Complex; previous GSWA site
10	125	0.4	11.4	Vasse Shelf sediments
9.9	42.1	0.1	5.5	Vasse Shelf sediments
18.7	147.2	0.4	15.2	Vasse Shelf sediments
75.8	275	11.7	40.1	Yallingup, Meelup National Park (MNP); Leeuwin Complex; regolith
12.6	168.0	6.2	15.7	Yallingup, Meelup National Park (MNP); Leeuwin Complex; granite outcrop #1
7.6	140.9	4.9	12.3	Yallingup, Meelup National Park (MNP); Leeuwin Complex; granite outcrop #2
20.3	82.5	4.4	11.5	Cowaramup, Vasse Shelf sediments
0.1	20.4	7.0	2.2	Meerlup Beach – Point Picouet

The RS-125 instrument (RSI, 2009) is calibrated, via "assay mode", to provide an approximation of U, Th and K concentrations in the half-space being sampled. The manufacturer indicates that an instrumental error of about 5 % is incurred in the RS-125 in the assay mode, due to the internally programmed calibration matrix. Variations from a perfect half-space can also cause some errors in the assay determination; observation locations were chosen to minimalise this type of error. The manufacturer's specifications of the RS-125 spectrometer indicate that the instrument reflects a sample volume of a radius of approximately one metre in the half-space. The Geiger Müller counter registers total counts with its design criteria focused largely at Uranium decay products. A statistically good and consistent relationship between the Geiger counter total counts and the RS-125 total counts was found (Middleton, 2013). Figure 4 of Middleton (2013) shows the correlation of the mean of the total counts between the RS-125 and Geiger Müller counter. A power relation was found: $y = 0.005x^{1.46}$, with a regression coefficient (R2) of 0.93, where y is total counts on the RS-125 instrument and x is total counts on the Geiger Müller counter.

Calibration procedures were carried out prior to the field acquisition with the RS-125. An important factor for calibration is the sampling time (or count accumulation time) of each measurement. In determining this, the manufacturer asserts that the calibration matrix, which is used to optimise the spectral definition of the three isotopes, is designed to provide the minimum error (ca. 5%) for sampling times of approximately 4 minutes. However, this has to be balanced against the requirement for longer sampling times for rocks with very low U, Th and K concentrations, as might be expected in many granites being investigated.

Repeat measurements were carried out on a granite slab from the Stoneville location (see Figure 5 for location) in the Darling Range for various sampling times in order to establish the optimum sampling time to return the most consistent U, Th and K "assay results" from the RS-125. Based on these measurements, a sampling time (count accumulation time) of 5 minutes was chosen for the field measurements for the RS-125. Similarly, 10 minutes was found appropriate for the Geiger Müller counter. For the sites visited during this study, at least three (n = 3), and often seven (n=7), repeat measurements were obtained at each location in order to obtain a basic understanding of possible statistical variation, due to the random nature of the decay of the radioisotopes in the granites. A larger number of repeats would have been desirable, but time limitations dictated the minimum of n=3.

A similar study was carried out in the Leeuwin Complex (see Figures 1 and 2 for general location). The mean, standard deviation and sample number (n) of the U, Th and K measurements for each site for Leeuwin Complex are shown in Table 2. The Leeuwin Complex and some new Darling Range data form new observational information presented in this review. These data contribute to our understanding of the radiogenic element abundance and heat generation capacity in Western Australia. The new data reflect expected similarities in thorium and uranium concentrations, and also correlate to higher than normal terrestrial heat generation.

Further, an analysis of RS-125 versus laboratory-based thorium and uranium was carried out for a number of the Leeuwin Complex measurement sites. Figure 7 shows this comparison. In most cases, measured the RS-125 and laboratory-based concentrations are fairly consistent. One significant reason for a slight difference in concentration is probably the sampling volume of the two methods. The RS-125 is sampling a half space (half-sphere) of radius of about one metre (see Figure 6), which contains about 5.55 x 10^3 kg of rock, versus a laboratory sample, which is comprised of only several kilograms of rock.

Relation of Airborne Radiometrics to Ground-based Measurements in Anomalous Thorium Regions

Recognised regions of anomalous thorium concentration considered in this review are (a) the Leeuwin Complex, (b) the Southern Perth Basin, (c) the Darling Range (Perth Hills) and parts og the Yilgarn Craton to the east, (d) the Northampton Complex, and (e) the Pilbara [Marble Bar-Wodgina] (see Figures 1, 2 and 11). Airborne radiometric data (and specifically thorium) are available for these regions (Brett, 2013).

With relation to the airborne radiometrics, various ground-truthing-verification studies have been carried out in the (1) Leeuwin Complex and southern Perth Basin, (2) Darling Range (Perth Hills) and Yilgarn Craton, (3) Northampton Complex and (4) Pilbara Region (Marble Bar-Wodgina region). These regions were selected for ground verification from previous studies, where aero-radiometric data indicated anomalously high concentrations of thorium and uranium (Middleton, 1912; Middleton, 1913; Middleton and Stevens, 2013; Brett, 1923; Sanders and McGuiness, 2000). For reasons mentioned above, it was suspected that these regions may possess actual surface thorium concentrations greater than the thorium threshold of 19.5 ppm derived from aero-radiometric observations.

In discussing the presence of thorium and uranium in near surface rocks, of which highly weathered components are referred to as "regolith", it is important to note that regolith does not in general constitute freshly outcropping rock. However, chemical elements, such as thorium and uranium, may be concentrated or diluted, depending on the processes that formed the regolith. Thorium is often considered not to be mobile (Mernagh and Miezitis,

2008), at least not as mobile as uranium, and as such probably concentrates during regolith-forming processes. Uranium is considered more mobile than thorium, and possibly becomes less concentrated during the regolith-forming process. This observation is consistent with thorium and uranium concentrations observed in the Darling Range, Yilgarn Craton and Leeuwin Complex, as often observed in this review.

Leeuwin Complex and Southern Perth Basin

The Leeuwin Complex is located in the south western part of Western Australia (Myers, 1990a; see Figures 1 and 2). The 2012 thorium map of Western Australia (Brett, 2013) shows a number of thorium anomalies with elevated concentrations in this region. Ground surveys described by Middleton and Stevens (2013) have shown thorium concentrations in this region to be very much greater than the threshold levels exhibited on the airborne-based maps. Table 2 shows these data, which exhibit thorium levels commonly greater than 150 ppm, while the upper threshold of airborne data is approximately 18 ppm.

Figure 8 shows thorium contours derived from the merge map of Brett (2013, and other sources), and eight ground measurement sites for the northern Leeuwin Complex and adjacent Perth Basin (Vasse region). The thorium concentrations at the ground measurement sites are shown in a table the figure. The highest thorium concentration occurs within a large anomaly near the town of Dunsborough. Ground measurements at locations 1 and 2, within this anomaly, yield thorium concentrations of 275 ppm and 168 ppm, respectively. These ground-based values are greater than the contour range of the anomaly 50 ppm to 148 ppm. Also another high concentration anomaly is seen on the western part of the map, and is identified as location 7 in Figure 8. A ground-based measurement at location 7 provides a thorium concentration of 147 ppm, which is consistent with the airborne-based thorium concentrations of between 100 ppm and 198 ppm shown in the contours.

Figure 9 shows a more detailed analysis of the anomaly near Dunsborough (locations 1 and 2 in Figure 8). In this analysis, two profiles (A-B and A-C) have been sampled across the thorium merge anomaly of Brett (2013). Location A on the profiles is situated in the ocean, offshore to the northwest of the Leeuwin Complex, and both profiles terminate at the southeastern end of the anomaly. Grey scale values (left-hand-side vertical axis) from the merge image are plotted against distance in pixel units (horizontal axis) in the profiles. The pixel distance of the profiles can be compared to horizontal distance in km in the map view also shown.

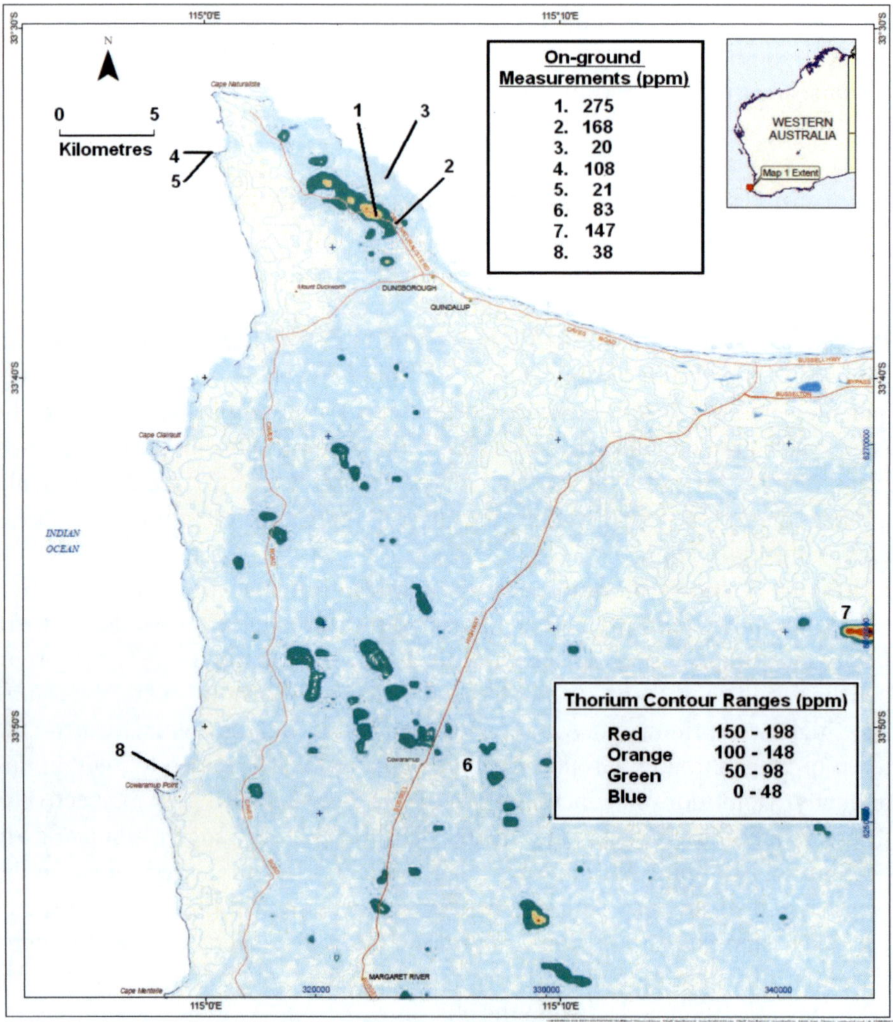

Figure 8. Map showing the Leeuwin Complex and adjacent Perth Basin in the southwest of Western Australia. The map shows locations of measurements, both laboratory samples and field-based samples [RS-125] (Middleton and Stevens, 2013). Locations 1 to 8 in the insert table show thorium concentrations in ppm. It is important to recognize that this region contains the highest known thorium concentrations in Western Australia. Source data from Cape Leeuwin 2011 Regional Airborne Survey (GA Project 1244) Thorium contours (derived May 2013; WACHEM data base).

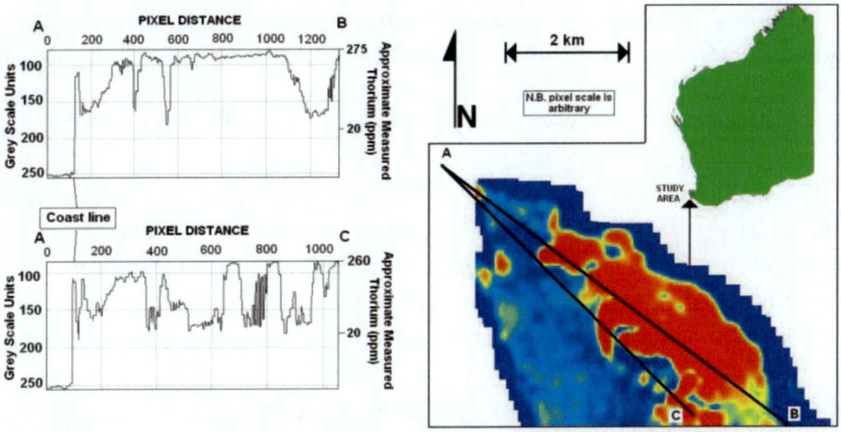

Figure 9. Profile of thorium concentrations across the Dunsborough thorium anomaly, based both on calibrated aero-radiometric thorium measurements (Brett, 2013) and field measurements (Middleton and Stevens, 2013) and newly acquired data for this paper. It is possible that the high values of thorium have been accumulated due to concentration with the development of the regolith in the region. Nevertheless, anomalously high thorium is also detected in fresh granitoid outcrops. This perhaps reflects a unique genesis of the Leeuwin Complex granitoids.

Further conversion from grey scale (although colour-scale converted) can be difficult to correlate to actual physical parameters. An approximate conversion of the grey scale values to ppm (parts per million) of thorium, based on surface measurements in the region (locations 1, 2 and 3 in Figure 8) is investigated with the Leeuwin data, and a correlation is not clear. The conversion from grey scale units to ppm thorium does not appear to be linear, although more surface data are required to establish an exact correspondence. Approximate conversion points can be pinned on (1) grey scale = 0 units, thorium concentration = 0 ppm, (2) grey scale = 175 units, thorium concentration = 20 ppm, and (3) grey scale = 120 units, thorium concentration = 275 ppm.

It should be noted that the highest thorium concentration (location 1 in Figure 8) of 275 ppm was situated on a layer about 15 to 0 m thick of surficial sediments possibly derived from underlying granitoid rocks (Lennard, 1991). The other ground-based thorium concentration measurement on this anomaly (location 2 in Figure 8) was situated directly on a granitoid outcrop, and indicated a thorium concentration of 168 ppm. The granitoids in this locality appear to have elevated thorium concentrations, but lower than the highest value (location 1) which suggest concentration of thorium due to chemical concentration from the process of chemical weathering. The profiles in Figure 9 indicate the highly variable nature of the surface thorium concentration across the anomaly. Further investigation is warranted into the geological and chemical processes entailed in the concentration of thorium in the Leeuwin Complex.

The Leeuwin Complex and adjacent Perth Basin probably constitute one of the most challenging areas of Western Australia to understanding of the distribution of thorium and its potential contribution to future resources, both in direct thorium supply and geothermal energy.

Darling Range (Perth Hills) and Yilgarn Craton

The Darling Range is a physiographic feature that typifies part of the western margin of the Yilgarn Craton in southwestern Western Australia. The Darling Range is an elevated region of about 300 to 400 m above the Perth coastal plain. The range is dominated by the Darling Fault, which is the geological boundary that separates the Yilgarn Craton from the Perth Basin (Cockbain, 1990; Playford et al., 1976). The long linear fault zone is one of the longest fault zones in the world, extending for about 1000 km (Figure 1). In the vicinity of Perth, the Darling Fault is expressed as a surface scarp of about 300 m, however, the subsurface expression of the fault zone on seismic reflection data (Iasky, 1993; Middleton et al., 1995) is observed to extend to over 10 seconds two-way time (approximately 50 km). It is not known if the same rocks that comprise the Yilgarn Craton underlie the Perth Basin, as it has been interpreted as a zone of major continental collision and orogenesis (Myers, 1990b).

Wilde et al. (1996) proposed that the western Yilgarn Craton has several component parts, and the region immediately to the east of Perth consists dominantly of geological units of the "Bailingup Terrane", of which the Perth Hills are comprised. Geological mapping in the Darling Range in the Perth region (Wilde and Low, 1978; Wilde and Low, 1980) indicates that regions of outcropping granitoids (felsic igneous rocks) are intersparced with rocks of mafic and ultramafic associations, and metasediments. The thorium appears to be associated with the rocks of granitoid association. Further, a Tertiary-aged lateritic profile of up to 100 m thick (Annand and Butt, 2003) is present over much of the region. The laterite profile was removed by erosion in a number of localities during the latter part of the Tertiary Period, generally as incised valleys, and in those places relatively fresh granitoids and mafic rocks are often exposed. Thorium values for granitoid, laterite and Tertiary channel locations are can be as high as in the granitoids themselves, and some values are reported in Table 3. Little work has been done to date on thorium concentrations in the laterite.

The Perth Hills lie in the central part of the western margin of the Yilgarn Craton (Figure 5). Knowledge of thorium concentrations in the Perth Hills region has been addressed by Middleton (2013), and new data reported since the initial study (Table 1). From the aero-radiometric mapping (RMA, 2010) and the ground measurements by Middleton (2013), the known thorium concentrations in this region vary between 0 and 55.7 ppm. The observed ground-based observations are often greater than the 19.5 ppm threshold set by the published aero-radiometric maps (RMA, 2010). This confirms that assertion that aero-radiometric-based estimates of thorium often require a ground-based reference.

Current projects by the Department of Mines and Petroleum are aimed to evaluate the maximum surface thorium concentrations in this region versus calibrated aero-radiometric observations. The aero-radiometric versus ground observations are also being correlated in a larger study of surface thorium and uranium concentrations within the whole Yilgarn Craton to the east of the Darling Range (Myers, 1990a; Myers, 1990b).

New field observations further east in the Yilgarn Craton, which are reported for the first time in this review, indicate that mean concentrations for thorium and uranium are 76.7 ppm and 10.4 ppm, respectively, for the locality of Doodlakine, and 65.2 ppm and 5.0 ppm, respectively, for the locality of Cunderdin (see Figure 2 for locations). While the Perth Hills are within the Yilgarn Craton, the Doodlakine and Cunderdin locations further to the east assumed some importance after a study by Jaeger (1970) showed that the Doodlakine locality exhibited higher than expected terrestrial heat generation and heat flow. Jaeger's conclusion of elevated heat generation at Doodlakine is due to the presence of elevated thorium and uranium concentrations in local granitoids. Jaeger (1970) used an equation similar to Equation 1 to estimate heat generation, where thorium and uranium concentrations were evaluated by laboratory-based techniques.

Northampton Complex

The Northampton Complex (see Figures 1 and 2 for location), which is part of the Pinjarra Orogen (Myers, 1990b), has been mapped on the

Geological Survey of Western Australia's Geraldton and Ajana 1:250,000 geological sheets (Hocking et al., 1982). The Northamption Complex forms the pre-sedimentary "basement" of the northern part of the Perth Basin, which is considered to possess appreciable geothermal energy potential, due to high temperatures observed within the sediments of the basin (Ghori, 2011). The high temperatures are presumed to be caused by elevated heat generation in the underlying Northampton Complex rocks.

Little recent geological work has been carried on the Northampton Complex. The Northampton Complex (see Figures 1 and 2 for location) has been mapped on the Geraldton and Ajana 1:250,000 geological map sheets (Hocking et al., 1982), and is part of the larger Pinjarra Orogen, which lies along the western margin of onshore Western Australia (GSWA, 1990; Myers, 1990b). On the Ajana 1:250,000 geological sheet, the Northamption Complex is dominated by rocks described as granulite and gneiss. These rocks have been metamorphosed under pressures in the range 600 to 900 MPa (6 – 9 kBar) and temperatures in the range of 600 to 800 °C (upper amphibolite to granulite facies metamorphic conditions; Sanders and McGuiness, 2000, and Turner, 1968). These rocks are not dissimilar to those of the Leeuwin Complex, which is also considered to be part of the Pinjarra Orogen (Myers, 1990b).

The merged thorium aero-radiometric map (Figures 2 and 4) suggests much of the Northampton Complex region consists of rocks with greater than 19.5 ppm thorium, noting that the proposed upper cut-off for the aero-radiometric maps for thorium is 19.5 ppm (RMA, 2010), as discussed above. However, also mentioned above and evident from ground-based studies in the Leeuwin Complex and Darling Range, one should be aware that 19.5 ppm appears to be the limit on the calibration technique of aero-radiometric observations for thorium, rather than a peak value that might be observed at the Earth's surface. Laboratory-based values of concentration reported by Sanders and McGuiness (2000) suggest some localities may be significantly higher (i.e. > 39 ppm) than the aero-radiometric threshold maximum of 19.5 ppm.

Sanders and McGuiness (2000) reviewed the geochemistry of the Ajana 1:250 000 geological sheet, and therein presented valuable data concerning the surface distribution of thorium in the regolith. For the purposes of this study, the regolith is defined as unconsolidated or indurated weathered rock, and includes residual and transported material that can cover and obscure underlying bedrock.

The thorium concentrations reported in this review are from the Geological Survey of Western Australia map of Sanders and McGuiness (Figure 10), and the values were measured by recognized laboratory methods referenced in the Sanders and McGuiness (2000) report. Their study presented a systematic grid of surface thorium concentrations in outcropping rocks at a spacing of approximately five to ten kilometres, nominally one sample per 16 km^2. The map (Figure 10) displays a range of thorium concentration between approximately 5 to > 39 ppm. Unfortunately, the upper limit of thorium concentration in the study, although acknowledged to be greater than 39 ppm, is not reported in their report.

An east-west profile across the Northamption Complex (Figure 12) based on the thorium merge map of Brett (2013) indicates that peak airborne-derived thorium concentrations are at the aero-radiometric threshold (17.92 ppm) for thorium. Therefore, this region is consistent with other regions of elevated thorium concentration in Western Australia that the aero-radiometric data can often represent lower thorium concentrations than appear at specific localities on the ground. Correlation of existing laboratory-based thorium concentrations (i.e. the Sanders and McGuiness, 2000 data) to RS-125 measurements of U, Th and K is planned for 2014-2015, and will be subsequently reported.

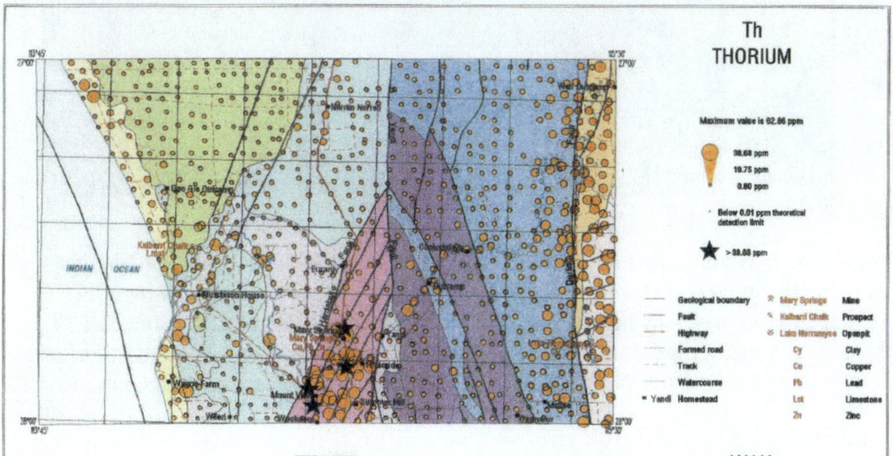

Figure 10. Map from Sanders and McGuiness (2000) showing the distribution of thorium concentrations measured in the regolith of the Ajana 1:250,000 geological sheet. The map shows variation in thorium from 5 to 39 ppm. Maximum values have not been reported, and no RS-125 comparison data are currently available.

Rocks from the Northamption Complex are very likely to extend under the north Perth Basin, and probably contribute to the high temperatures observed in petroleum wells in that region (Ghori, 2008; Mory and Iasky, 1996). Little definitive work has been carried out to date on the influence of radiogenic elements in the basement rocks upon elevated temperatures in the sediments of the north Perth Basin.

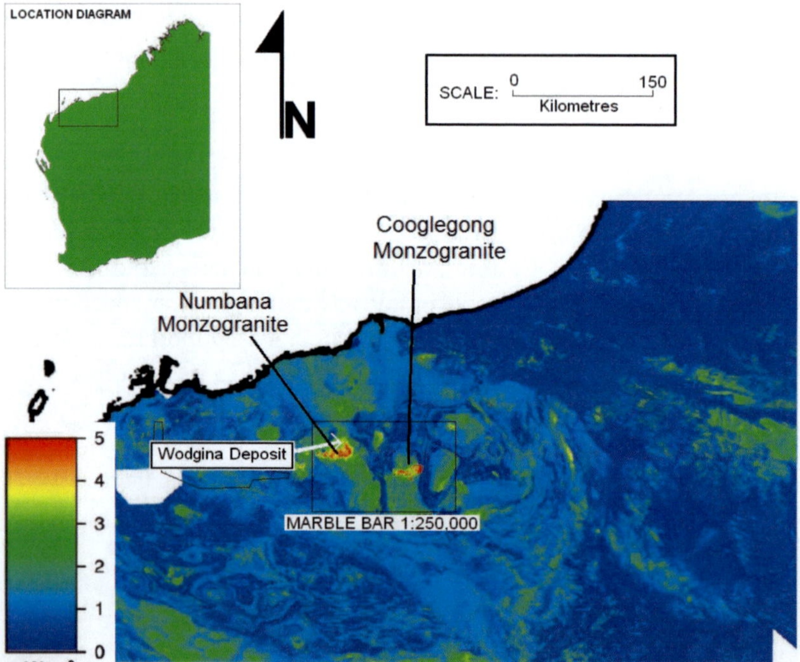

Figure 11. Map showing high heat generation localities in the northern Pilbara region, epecifically the Marble Bar 1:250 000 geological map sheet (Western Australia Geological Survey). The heat generation map was derived from Th, U and K concentrations via Equation 1. The high heat generation (and consequently thorium) areas occur in the region of the Numbana Monzogranite, and are described in the text.

Pilbara (Marble Bar-Wodgina)

The Pilbara region of Western Australia is well known for its richness in iron ore, copper lead zinc and gold. Other than these minerals, tantalum is mined in the Wodgina region near Marble Bar (Figure 11 shows location). The Numbana Monzogranite, which is located on the Marble Bar 1:250 000 geological sheet (Geological Survey of Western Australia), shows a distinct thorium and uranium anomaly. This anomaly exhibits a close association with the Wodgina tantalum deposit. Data from the Geological Survey of Western Australia database

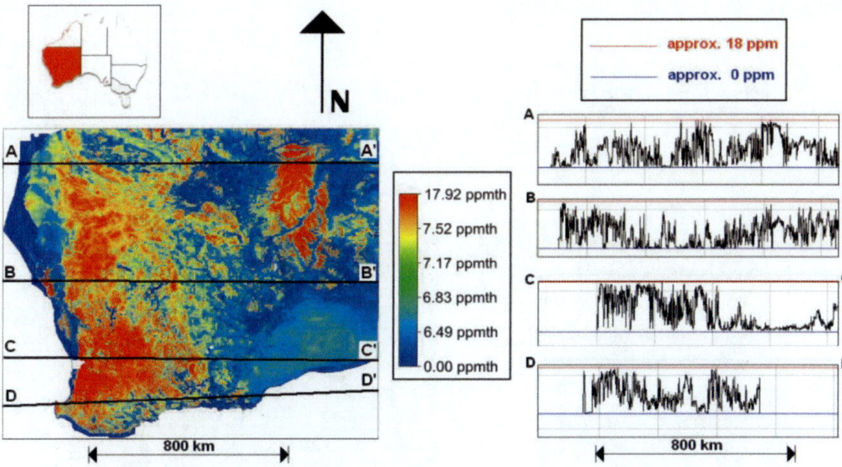

Figure 12. Profiles of thorium concentrations, based on calibrated areal thorium measurements. The four profiles cover (A – A') the Pilbara region, (B – B') the Northamption region into the Officer Basin, (C – C') the Darling Range in the vicinity of the Perth Hills to near Ceduna, and (D – D') the Leeuwin Complex and southern coast of Western Australia. The profiles are limited by the "cut-offs" of the merge process of the diversely sourced airborne radiometric data (RMA, 2010; Brett, 2013). An upper cut-off is taken as 17.92 ppm and the lower cut-off is zero. It should be noted that the while the northern part of the Leeuwin Complex (western end of profile D – D') contains the highest ground-based thorium concentrations (275 ppm) recorded in studies reported in this review to date in Western Australia, this is not reflected in these profiles, because of the averaging of the pixel size.

The Wodgina 1:100,000 geological sheet in the Pilbara region of Western Australia exhibits outcrops of the Numbana Monzogranite (Blewett and Champion, 2005; Hickman, 2012). This rock unit was highlighted by Middleton (2012) as possessing sufficiently high levels of thorium and uranium to provide a crustal geothermal energy source. The thorium merge map by Brett (2013) shows elevated thorium in the Wodgina, and nearby localities in the Marble Bar region, as identified by Middleton (2012) and previous investigators (see Figure 4). Middleton (2012) proposed a correlation of high radiogenic granites to magnetic and gravity anomalies in this region. However, this correlation may be due to other petrophysical properties, such as density and magnetic susceptibility, of the Numbana Monzogranite than the presence of thorium. Nevertheless, petrophysical properties, such as density and magnetic susceptibility, which define the Numbana Monzogranite may also possess an associated thorium signature. There is strong support that the gravity, magnetic and thorium data sets for the Numbana Monzogranite exhibit strong spatial correlation.

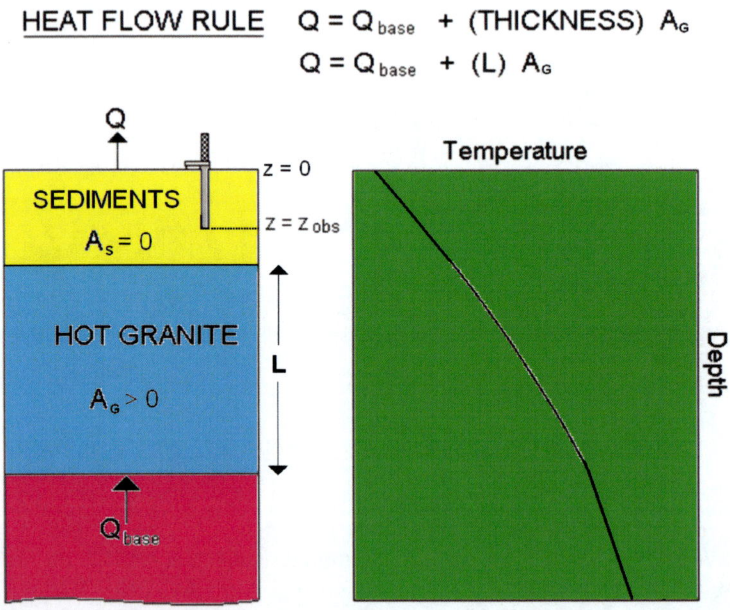

Figure 13. A schematic model showing the surface heat flow (Q_s) from a layered Earth comprised of a layer sediments with no heat generation overlying a radiogenic - granitoid layer of thickness L, and deeper layer with heat flow of Q_b (or Q_{base}). These are the spatial assumptions for the temperature calculation in Equation 3.

A map of heat generation in Western Australia was generated by Middleton (2011), based on Equation 1. The map reflects the input of thorium and uranium, and to a lesser extent potassium, to heat generation of surface geology. This map is indicative of high heat generation, and also high Th and U content, as expressed in Equation 1. Two particularly anomalous regions of high-radiogenic heat generation are shown in Figure 11, which is a sub-region of the original state map. The two regions are identified to be high in radiogenic elements (i.e. Th and U), and these are within the Numbana and Coolegong monzogranite intrusions (Figure 11). These intrusions, which crop-out, are late-stage, highly fractionated granites (Van Krandendonk, 2011) and known to possess high U and Th concentrations.

Thorium Overview

An overview of the thorium distribution in the southern part of Western Australia is valuable in the light of limited infrastructure within the state. Regions to the north of the State of Western Australia possess limited infrastructure, with respect to roads, railways and associated energy supply. Localities in the south of the State have greater advantage for short-term commercial development, however, some localities in the Pilbara region (see Figure 1 for location) also possess infrastructure advantage, due to long-term mining and petroleum developments.

Four cross-sections of thorium concentration, based on the merge maps of Brett (2013), are shown in Figure 12. These cross-sections are located in southern-central of Western Australia, where infrastructure in likely to be better than the northern regions of the state. The cross-sections show thorium concentration variability with in a fairly consistent fashion across the State, although maxima in thorium concentrations are not well represented in the merged aero-radiometric survey data.

The cross-sections show relative thorium concentration on a grey scale from the thorium merge maps of Brett (2013) versus pixel distance (true distance can be determined from the associated map). The cross-sections are limited by the maximum aero-radiometric-calibrated thorium value of 17.92 ppm. Approximate maximum and minimum concentrations are shown also, but these are based on the thresholds of the aero-radiometric thorium calibrations (0 to 17.92 ppm), as opposed to calibrated ground-truthing values. It is important to realize that insufficient ground-based thorium concentrations are available to correlate the airborne radiometric data to "ground-truthed" thorium concentrations at the present time. In fact, it may not be possible to reconcile the two data sets, airborne versus ground-based, due to the data gathering procedures, but some effort needs to be made to reconcile these differences.

THORIUM IN A GEOTHERMAL CONTEXT IN WESTERN AUSTRALIA

General Background to Heat Flow and Heat Generation

It is commonly recognised that granitoids can contain a high content of the radiogenic elements Uranium (U), Thorium (Th) and Potassium (K). These rocks have often been referred to as "hot or radiogenic granites", and provide a dominant component to the heat flow in the Earth. Two geothermal regimes are recognised in Australia, and they are the hot dry rock (HDR) and the hot sedimentary aquifer (HSA) regimes. In HDR regimes, hot granites reside beneath a thermal blanket comprised of between three to five kilometres of sedimentary rocks, and water is injected into fractured hot granites to provide a source of hot water for geothermal energy. In HSA regimes, likewise hot granites reside beneath a sedimentary cover, but the natural hot water in the sediments, caused by heat generation in the deeper hot granites, is extracted for geothermal energy. A theoretical review of temperature, heat flow, heat generation and constituent radiogenic elements follows.

Heat flow at the surface of the Earth (Q_s), over a series of n layers with internal heat generation and stable tectonic environment, can be determined in a 1D case by the equation (Carslaw and Jaeger, 1959; Garland, 1971; Kappelmeyer and Haenel, 1974):

$$Q_s = Q_b + A_1H_1 + A_2H_2 + ... + A_nH_n, \qquad (2)$$

where Q_b is the heat flow from the base of the n layers being considered, A_1 is the heat generation in layer1, H_1 is the thickness of layer 1, A_2 is the heat generation in layer 2, H_2 is the thickness of layer 2, A_n is the heat generation in layer n, H_n is the thickness of layer n.

Temperature within a surface layer of thickness of L, and uniform heat generation of A_o, is shown by Carslaw and Jaeger (1959) to be described by (see Figure 13):

$$T = \left[\frac{A_o z}{2K}\right](2L - z) + \frac{Q_b z}{K} + T_s, \qquad (3)$$

where T is temperature, z is depth of observation ($0 \le z \le L$), K is thermal conductivity, T_s is mean annual surface temperature, and Q_b is the heat flow at the base of the layer. This is the large time (t->∞) solution of the 1D heat conduction equation. The geometry applicable to this equation is shown in Figure 13, and equation (3) can be applied to estimate the temperature versus depth in a granitoid body with uniform radiogenic heat generation after a very long residence time. Shorter residence times for a time-transient thermal source is dealt with below.

Modeling can be undertaken, using Equation 3) to understand temperatures in the Earth's crust near to the surface (within 3 to 5 km). This is largely driven by a desire to understand geothermal potential and current temperatures in the near surface in both the crystalline and sedimentary blanketed crust in Western Australia. The pertinent unknown parameters in this modelling exercise are the assumed heat generation within the surface layer of hot granite (A_1) and its thickness (L). The other parameters to be introduced into the model are relatively well known, these being thermal conductivity (K), thermal diffusivity (α), background heat flow (Q_b) and surface temperature (T_s).

The large-time (t->∞) solution, Equation 3 above, may not be valid in the case of recent geological activity. It should be recognised that differences of up to several °C from the large-time equation (Equation 3), i.e. t -> ∞, when a time-transient model is applicable.

In terms of a more complete model for dynamic geological processes, a fuller solution is proposed by Carslaw and Jaeger (1959, p. 97) for the case of a surface layer of finite thickness with uniform heat generation within it, and a limited time since the presence of the heat generation, is:

$$T = \left[\frac{\alpha A_o z}{K}\right]\left[1 - 4i^2 erfc\left(\frac{z}{\tau}\right) + 2i^2 erfc\left(\frac{L+z}{\tau}\right) - 2i^2 erfc\left(\frac{L-z}{\tau}\right)\right] + T_o erf\left(\frac{z}{\tau}\right) + \frac{Q_b z}{K}, \quad (4)$$

where

T: temperature,
z: depth,
t: time,
α: thermal diffusivity,
K: thermal conductivity,
A_o: heat generation,
T_o: initial temperature of layer at t = 0,
Q_b: heat flow at the base of layer,
erf (x): error function

erfc(x): complimentary error function [1 – erf(x)]
i^2erfc(x): 2nd integration of the error function, and
$\tau = \sqrt{4\alpha t}$.

Equations 3 and 4 is used in modeling results for various sites in Western Australia below. The geometry of the 1D model used in the case of a sedimentary layer is shown in Figure 13, and in the case of no sedimentary layer is shown in Figure 14. In both cases, the thickness of the radiogenic-rich layer is designated by L.

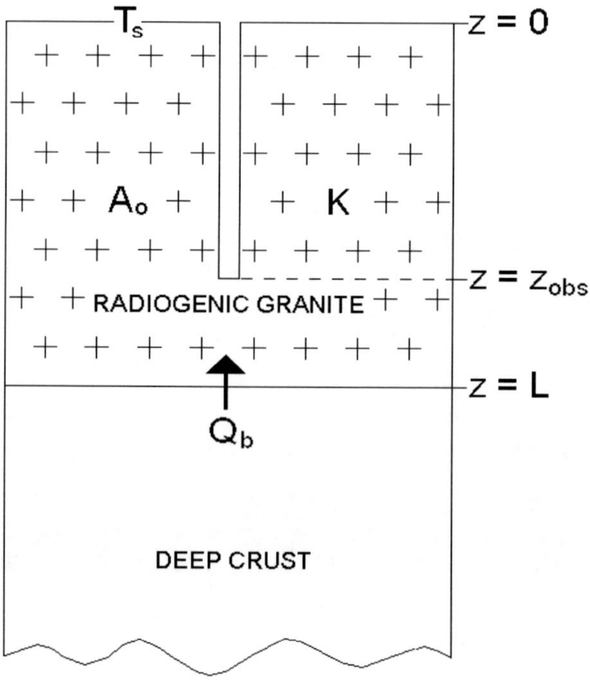

Figure 14. A schematic model showing the model proposed by Jaeger (1970) for the crustal structure of the Yilgarn Craton, based on heat flow and heat generation measurements. This has not recently changed in recent times, although the Darling Range data may suggest an increase in heat generation to the west of the Yilgarn Craton.

In reference to geothermal modeling, it is of value to reinforce that A_0 is strongly related to thorium concentration, C_{th}, which in many cases in Western Australia is greater than the usually expected uranium concentration level, as indicated in Equation 1. If Th/U > 3.7 (see below), then thorium will contribute more to radiogenic heat generation than uranium. Terrestrial heat generation and heat flow in regions, such as the Leeuwin and Northampton complexes in Western Australia, together with some localities in the Yilgarn Craton (for example, specific measured locations being Doodlakine and Cunderdin; locations in Figure 2, and values in Tables 1 and 3), is often dominated by thorium concentration.

Role of Thorium in Terrestrial Heat Generation

In a geothermal context, and especially in terms of heat generation (Kappelmeyer and Haenel, 1974), the relative concentrations of uranium (U), thorium (Th) and potassium (K) in the constituent granitoids within the relevant crustal units provide a dominating heat generating influence.

Radiogenic heat generation (A_o) derived from the mean U, Th and K elemental concentrations, using Equation 1, based on Kappelmeyer and Haenel (1974) to convert U, Th and K concentrations into heat generation, are also shown in Tables 1 and 2. The factors are based on theoretical determinations of the heat created by the decay products of each of the elemental isotopes being absorbed into the host rock. The factors used in this study are, 0.26 $\mu W\ m^{-3}$ for U concentration in parts per million (ppm), 0.07 $\mu W\ m^{-3}$ for Th concentration in ppm, and 0.10 $\mu W\ m^{-3}$ for K concentration in percent, and are consistent with those published elsewhere in the literature (Beardsmore and Cull, 2001; Bradey et al., 2006; Jessop, 1990; Joshua et al., 2008).

Table 1 from Bradley et al. (2006) provides support for the proposal that one ppm thorium concentration can contribute 0.069 $\mu W\ m^{-3}$, and one ppm uranium can contribute 0.26 $\mu W\ m^{-3}$, to heat production in a body of density about 2690 $kg\ m^{-3}$ within the Earth's crust. Clauser (2009; 2011, Table 2) indicates that thorium may contribute between 25.6 to 27.0 $\mu W\ kg^{-1}$, or approximately 0.068 $\mu W\ m^{-3}$ (assuming an average rock density of 2650 $kg\ m^{-3}$), to crustal heat production. Jaeger (1970) used factors of 0.2 $cal\ g^{-1}\ yr^{-1}$ for each part per million (ppm) of thorium, 0.73 $cal\ g^{-1}\ yr^{-1}$ for each ppm of uranium, and 27×10^{-6} $cal\ g^{-1}\ yr^{-1}$ for each percent of potassium in a rock. These factors convert approximately to 0.084 $\mu W\ m^{-3}$ for thorium, and 0.305 $\mu W\ m^{-3}$ for uranium. These studies support the assumption that the value of 0.07 $\mu W\ m^{-3}$, proposed herein, to convert thorium concentration to radiogenic heat generation is not unreasonable.

Therefore, it is assumed for this review that heat generation, A_o in units of $\mu W\ m^{-3}$, is determined by (which is the same as Equation 1):

$$A_o = 0.26\ C_U + 0.07\ C_{Th} + 0.10\ C_K\ , \qquad (5)$$

where C_U is the concentration of uranium in ppm, C_{Th} is the concentration of thorium in ppm, and C_K is the concentration of potassium in percent. Alternatively, an approach to estimating near-surface heat generation is to use equation 5 to estimate U, Th and K concentrations from calibrated airborne radiometric data.

A heat generation map, based on airborne radiometric data, was published by Middleton (2011, Figure 1). This type of map is based on U, Th and K airborne-radiometric data and maps released by Geoscience Australia (Minty et al., 2009; Stoltz, 2010; Budd et al., 2011). Essentially, these data represent an approximation of the average surface heat generation for all the rocks in each pixel on the map. In this context, Figure 7 of Middleton (2013) shows a surface heat generation map, which has pixel dimensions of approximately 1.3 km x 1.3 km. However, it should be recognised that such an average of heat generation may not be representative of any particular outcrop within the pixel area of this map. Therefore, while such maps are a good general guide to high heat generation regions, the actual heat generation values on the map may be significantly different than those found for specific geological units on the ground within the pixel area. For Example, the Darling Range appears to have a maximum surface radiometric heat generation in the range of 2.5 to 5.0 $\mu W\ m^{-3}$ (Middleton, 2013: Figure 7), which is lower than many of the observed field measurements of the outcropping granites in the region (see below, Tables 1 & 3). Such results might not be unexpected, due to the averaged nature of radiometric data, and have been discussed above with respect to thorium concentration.

Influence of Thorium for Heat Generation and Heat Flow in WA

Doodlakine Revisited

The Doodlakine site (see Figure 2 for location) was considered to possess unusually high heat generation and heat flow by Jaeger (1970). This site was investigated with field observations, using the RS-125 instrument in early 2014, these results are reported for the first time in this review (Tables 1 and 3). The recent investigation with the RS-125 instrument found that mean concentrations for thorium of 76.7 ppm, uranium of 10.4 ppm and potassium of 4.0 % were found in outcropping granitoids. These combined U, Th and K concentrations indicate, through Equation 5, a local heat generation of 8.5 μW m^{-3}, which is consistent with the heat generation reported by Jaeger (1970). Jaeger (1970) calculated a heat generation from surface granitoid samples of 9.2 μW m^{-3} using different elemental conversions from the values adopted in Equation 5. If the elemental conversion values used by Jaeger (0.084 for Th and 0.305 for U) are applied to the Doodlakine elemental values above, a heat generation of 9.6 μW m^{-3} is found, which is within 5% of the value recently measured and used in this review.

Broader Heat Generation and Heat Flow Studies

A wide range of surface heat flow values (Sass et al., 1976) are predicted for radiogenic granites from both observation and modeling, and is expected in the range ($6 < Q_s < 31$ mW m^{-2}). Surface heat flow may be diagnostic of radiogenitic-granitoid thickness, if the assumption of almost uniform heat generation within the granitoid body is correct. A modification of Equation 2 can be made in the case of a single uniform radiogenic body in the crust, where surface heat flow (Q_s) is related to basal heat flow from the mantle (Q_b), heat generation in the radiogenic body (A_0) and the thickness of this body (L):

$$Q_s = Q_b + A_o L \tag{6}$$

Equation 6 was used by Jaeger (1970) to estimate the thickness of the radiogenic layer (L) and the basal heat flow (Q_b) from measured values of surface heat flow (Q_s) and heat generation (A_o) within the granitoids of the Yilgarn Craton.

The work by Jaeger (1970) suggested a thickness (L) of 4.5 km for the radiogenic Crust (i.e. the part of the Crust providing significant radiogenic heat input to the Earth's heat flow), and a basal heat flow (Q_b) of 26.3 mW m^{-2} to Yilgarn Craton (1D model shown in Figure 14). This type of heat generation and heat flow model covers a substantial portion of the state where there is no substantial sedimentary cover. Very few data are available on the variation of heat generation within the granitoid bodies in the Yilgarn Craton, and more work is needed in this area. However, apart from the uncertainty in the uniformity of heat generation within heat generating bodies, a major uncertainly remains to be the thickness of the granitoid bodies in the various geological provinces in Western Australia. Without further drilling, geophysical surveying and interpretation of existing geophysical data, this uncertainty will remain.Little work on this has been done with regard to heat generation and heat flow determinations in Western Australia since.

In a broader context, Hot Dry Rocks (2008), Ghori (2008), and Gibson and Ghori (2011) have reviewed temperatures and heat flow in the Perth Basin. These studies suggest that in some places the basal heat flow into the Perth Basin sedimentary pile typically falls in the range of 40 to 120 mW m^{-2}. Middleton (2010) found typical surface heat flow in the northern Perth Basin to be about 85 mW m^{-2}, but with some localities as high as 130 mW m^{-2}, which is consistent with the heat generation regime found in the Northamption Complex, discussed above. Given that heat flow in the radiogenic granites of the Darling Range from modelling by Middleton (2011 and 2012) varies from 34 to 96 mW m^{-2}, it is not unreasonable to conclude that the higher levels of heat flow in the Perth Basin may be caused by the presence of radiogenic granites in the basement rocks, with similar associations to the Perth Hills and Yilgarn Craton (Myers, 1990b; Iasky, 1993).

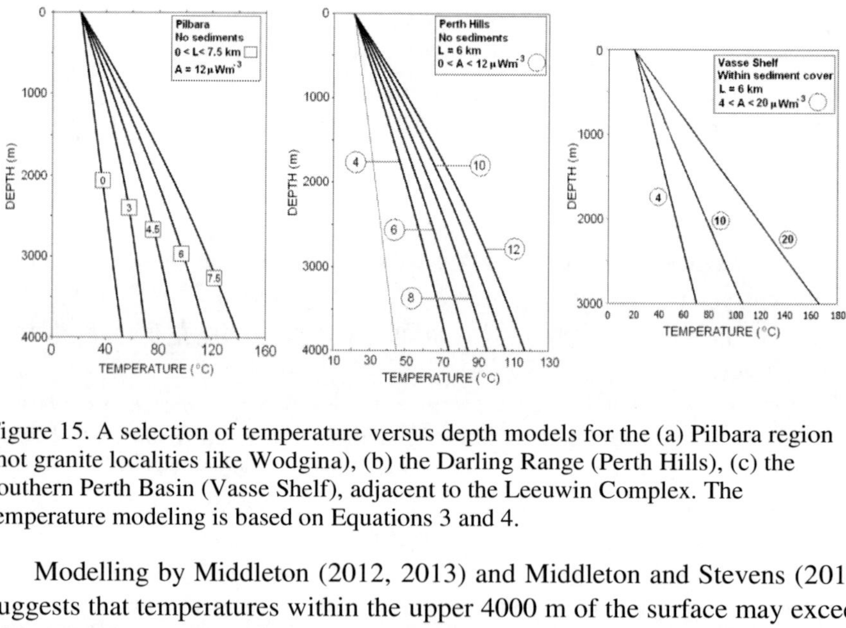

Figure 15. A selection of temperature versus depth models for the (a) Pilbara region (hot granite localities like Wodgina), (b) the Darling Range (Perth Hills), (c) the southern Perth Basin (Vasse Shelf), adjacent to the Leeuwin Complex. The temperature modeling is based on Equations 3 and 4.

Modelling by Middleton (2012, 2013) and Middleton and Stevens (2013) suggests that temperatures within the upper 4000 m of the surface may exceed 100 °C (Figure 15). It is shown that temperatures within radiogenic-rich granitoid bodies may reach a maximum of about 110 ºC at depth of 4 km in the Darling Range. The temperature within sediments, adjacent to the Leeuwin Complex in the southern Perth Basin, may reach temperatures of over 130 ° C at a depth of 3 km.

Geothermal Commercial Considerations in Australia

In consideration of commercial geothermal energy production in the region, the modelled temperatures cannot support current concepts of large-scale electricity generation with geothermal energy based on the *Hot-Dry-Rock-type* scenario, where it is generally accepted that temperatures well in excess of 220 ºC are required for electricity generation. However, the new generation of low-temperature, organic-Rankine-cycle turbines (Welch and Boyle, 2009; Welch et al., 2010), and adsorption chillers, which operate in the vicinity of 70 to 120 ºC, could place the Darling Range in an advantageous economic perspective for alternative energy supply close to an existing market.

This study suggests that realistic temperatures between depths of 500 m and 1500 m, which are realistically drillable depths, in the Vasse region may vary between 30 °C and 45 °C. This temperature range is confirmed by hydrogeological drilling. If a hot granitoid is particularly thick (thickness about 6 km) then a temperature of about 60 °C may occur at 1500 m. The observed data and modelling clearly indicates that the Vasse region is a low-temperature geothermal regime. Unless wells are drilled to depths of greater than 3000 m, electricity generation is not an option. Nevertheless, shallow wells or boreholes can replicate the hot springs scenario of the Peninsular Hot Springs resort in Victoria, and this is not an insignificant industry when in a high tourist destination, such as in the Margaret River region in the southwest of Western Australia. Similarly, geothermal regimes dominated by *Hot-Sedimentary-Aquifer* systems (Ghori, 2008), as the Perth metropolitan area, where heated swimming pools are an active geothermal enterprise.

Geothermal-electricity generation, albeit on a small community scale like (Birdsville in Queensland or Margaret River in Western Australia), will require deeper drilling. The modelling suggests that wells of 1700 m to 2700 m can fulfil the requirement of the 100 °C groundwater that supplies the Birdsville geothermal plant. The answer that is not currently known is how many, and what volume of, granites residing deep in the Vasse region possess the necessary high heat generation of 10 to 20 μW m^{-3}; this level of natural heat generation is required to provide geothermal heat to sustain the region's energy. The 100 °C maximum temperature range is also sufficient to permit electricity generation with organic rankine cycle (ORC) or variable phase cycle (VPC) turbines (Welch and Boyle, 2009).

Table 3. Thorium and uranium concentration for various locations used in this review, and the thorium/uranium ratio. Yilgarn refers to the Yilgarn Craton, Pilbara refers to the Pilbara Craton, Leeuwin refers to the Leeuwin Complex and the Perth Basin is as such. The term (felsic) refers to granitoid rocks, mafic to rocks rich in magnesium and iron, and laterite and sediments are as such. Regolith/felsic refers to a layer of weathered rock overlying granitoids. It should be noted that a thorium/uranium ratio greater than 3.7 indicates that thorium has a greater input into heat generation than uranium. Thorium/uranium ratios greater than 10 appear to be rare in the Yilgarn Craton, but common in the Leeuwin Complex. Thorium/uranium ratios greater than 10 may reflect a prevalence of uranium depletion in those specific localities

LOCATION	THORIUM (ppm)	URANIUM (ppm)	Th/U (ratio)	COMMENTS
Stoneville 1	32.6	6.9	4.7	Yilgarn (felsic)
Stoneville 2	35	16	2.2	Yilgarn (felsic)
Ashendon Rd	44.7	7.3	6.1	Yilgarn (felsic)
Golden View 1	36.8	11	3.3	Yilgarn (felsic)
Golden View 2	42.1	9	4.7	Yilgarn (felsic)
Golden View 3	33.6	6.7	5.0	Yilgarn (felsic)
Darlington	20.7	6.1	3.4	Yilgarn (felsic)
Canning Dam	30.2	7.3	4.1	Yilgarn (felsic)
Kalamunda East	55.7	22.7	2.5	Yilgarn (felsic)
Lesmurdie Falls	54.3	5.7	9.5	Yilgarn (felsic)
Glen Forrest 1	45.0	17.0	2.6	Yilgarn (felsic)
Glen Forrest 2	42.6	10.1	4.2	Yilgarn (felsic)
Glen Forrest 3	40.1	15.8	2.5	Yilgarn (felsic)
Glen Forrest 4	39.2	13.0	3.0	Yilgarn (felsic)
Mundaring Weir 1	34.4	6.1	5.6	Yilgarn (felsic)
Mundaring Weir 2	31.2	3.6	8.7	Yilgarn (felsic)
Mundaring 1	7.6	1.5	5.1	Yilgarn (Tertiary channel)
Mundaring 2	68.6	6.0	11.4	Yilgarn (laterite)
Parkerville	42.2	9.4	4.5	Yilgarn (felsic)
Doodlakine	76.7	10.4	7.4	Yilgarn (felsic)
Cunderdin	65.2	5.0	13.0	Yilgarn (felsic)
Numbana	61.3	31.3	2.0	Pilbara (felsic)
Coolegong	48.7	19.3	2.5	Pilbara (felsic)

LOCATION	THORIUM (ppm)	URANIUM (ppm)	Th/U (ratio)	COMMENTS
Sugarloaf Rock 1	108.4	6.0	18.1	Leeuwin (felsic)
Sugarloaf Rock 2	21.3	1.3	16.4	Leeuwin (mafic)
Gracetown North	37.9	2.1	18.0	Leeuwin (felsic)
Gracetown South	24.4	1.8	13.6	Leeuwin (felsic)
Leeuwin Lighthouse	37.8	2.8	13.5	Leeuwin (felsic)
Water Wheel	46.8	3.3	14.2	Leeuwin (felsic)
Brown Hill Estate	125.0	10.0	12.4	Perth Basin (sediments)
Gale Road Quarry	42.1	9.9	4.3	Perth Basin (sediments)
Farquhar Road	147.2	18.7	7.9	Perth Basin (sediments)
Meelup Nat. Park 1	275.0	75.8	3.6	Leeuwin (regolith /felsic)
Meelup Nat. Park 2	168.0	12.6	13.3	Leeuwin (felsic)
Meelup Nat. Park 3	140.9	7.6	18.5	Leeuwin (felsic)
Meelup Beach	20.4	0.1	204.0	Leeuwin (felsic)
Cowaramup	82.5	20.3	4.1	Leeuwin (felsic)

THORIUM - URANIUM RATIO IN WA

A fair volume of work has been carried out during heat generation studies to document thorium and uranium abundances in Western Australia. Table 3 shows thorium and uranium concentrations and thorium/uranium (Th/U) ratios acquired for a number of the studies reported in this review.

This ratio is quite variable. Nevertheless, the ratio is fairly consistent in granitoids in the Yilgarn and Pilbara cratons. The exception is in the Leeuwin Complex and adjacent Perth Basin (Vasse Shelf) in the southwest of Western Australia. It is uncertain if these thorium and uranium concentrations are specifically related to the tectonogenesis of the Leeuwin Complex (Pinjara Orogen; Myers, 1990), regolith formation and associated uranium mobility during weathering processes (Annand and Butt, 2003; Mernagh and Miezitis, 2008), or some other currently unknown cause.

It is, however, recognized that Th/U > 3.7, based on Equation 1, indicates that thorium has a greater input into the heat generation of a rock than uranium. This situation occurs often in Western Australia, and is probably more the rule than the exception, as has been documented above. It is not fully clear why the Th/U ratio can vary up to 15 (see Table 3), but this variability is currently assumed to be due to greater uranium mobility versus thorium during secondary weathering and erosion processes (largely during the Tertiary Period; Annand and Butt, 2003). A question that can be reasonably posed: Is uranium depletion systematically related to thorium concentration over an equivalent geological timeframe?

Conclusion

A broad understanding of the distribution of thorium resources in Western Australia is already known, but detailed distributions are sketchy. The potential contribution to the World's economic requirements for thorium is largely unknown. Impacts of thorium distribution on geothermal projects in Western Australia could be substantial, especially in the south west of the state, where thorium is known to be anomalously high. Areas of particular interest are in the Vasse Shelf region of the southern Perth Basin, adjacent to the Leeuwin Complex, and in the north Perth Basin, where high-thorium-concentration basement rocks of the Northampton Complex underlie this part of the Perth Basin.

Elsewhere in Western Australia, thorium concentrations are known to be moderately high (within the threshold boundary of about 18 ppm) at a regional levels from the aero-radiometric data, but little is known about localized concentrations. The thorium volumes in Australia, proposed by Mernagh and Miezitis (2008), may require further review for a global comparison. The refinement of thorium, distributed at such low concentrations (often < 100 ppm) may provide an obstacle, but perhaps not an insurmountable barrier, for the economic development of thorium in Western Australia especially as knowledge of the distribution of local thorium concentration grows.

Nevertheless, the anomalous thorium concentration levels in the Leeuwin Complex in the south-west of Western Australia may provide an economic motive for both geothermal development and thorium elemental extraction. The anomalously high thorium concentrations observed in the Leeuwin

Complex provide the greatest scientific challenge to understand the distribution of thorium in Western Australia.

ACKNOWLEDGMENTS

The Department of Mines and Petroleum (DMP) in Western Australia is acknowledged for supporting this review paper, and permission to publish by the Executive Director of the Petroleum Division, Jeff Haworth. Valuable contributions were made by Mark K. Stevens, Senior Geologist from the Resources Branch of the Petroleum Division (DMP). Arthur Mory, Senior Geologist (albeit senior curmudgeon, with the nicest possible inference) from the Geological Survey Division (DMP), also provided some well considered comments.

REFERENCES AND BIBLIOGRAPHY

Annand, R.R., and Butt, C.R.M., 2003, Distribution and evolution of 'laterites' and lateritic weathering profiles, Darling Range, Western Australia, *Australian Geomechanics*, v. 36, No. 4, p. 41-58.

Appleyard, S.J., 1991, The geology and hydrogeology of the Cowaramup boreline, Perth Basin, Western Australia, *Geological Survey of Western Australia,* Hydrological Report 30.

Baxter, J. L., 1977, *Heavy mineral sand deposits of Western Australia:* Western Australia Geological Survey, Mineral Resources Bulletin 10.

Beardsmore, G.R. and Cull, J.P., 2001, *Crustal Heat Flow: A guide to Measurement and Modelling,* Cambridge University Press.

Blewett, R.S. and Champion, D.C., 2005, *Geology of the Wodgina 1:100 000 sheet,* Western Australia Geological Survey 1:100 000 Geological Series Explanatory Notes.

Boyle, R.W., 1982, *Geochemical prospecting for thorium and uranium deposits,* Elsevier Scientific Publishing Company, Amsterdam, 498p.

Brady, R.J., Ducea, M.N., Kidder, S.B., and Saleeby, J.B., 2006, The distribution of radiogenic heat production as a function of depth in the Sierra Nebada Granitic body, California, *Lithos*, 86, 229-244.

Brett, J, 2013, T*horium merge map of WA, Geological Survey of Western Australia,* electronic publication.

Budd, A.R., Gerner, E., Meixner, A.J., Kirkby, A.I., Weber, R., and Spring, J., 2011, Geoscience Australia's onshore energy security program: geothermal energy, in: Budd, A.R. (editor), 201, *Proceedings of the 2011 Australian Geothermal Energy Conference*, 16-18 November, Melbourne, GeoscienceAustralia, Record 2011/43,31-36.

Carslaw, H.S. and Jaeger, J.C., 1959, *Conduction of heat in solids*, Oxford University Press.

Clauser, C., 2009, Heat transport processes in the Earth's crust, *Surveys in geophysics,* 30, 163-191.

Clauser, C., 2011, Radiogenic Heat Production of Rocks. *In* Harsh Gupta (Ed.), *Encyclopedia of Solid Earth Geophysics*, 2nd edition, Springer, Dordrecht.

Cockburn, A.E., Hocking, R.M. and Myers, J.S., 1990, Geological Evolution, Phanerozoic, in *Geology and Mineral Resources of Western Australia*, Western Australia Geological Survey, Memoir *3,* 735-755.

Garland, G. D., 1971, *Introduction to Geophysics, mantle, Core and Crust*, W.B. Sanders, Toronto, 420p.

Ghori, K.A.R., 2008, Perth Basin's geothermal resources, *Australian Geothermal Energy Conference 2008, Extended Abstracts*, 55-61.

Ghori, K.A.R., and Gibson, H., 2011, Geothermal simulation, Perth Basin, Western Australia, in: Budd, A.R. (editor), 2011, *Proceedings of the 2011 Australian Geothermal Energy Conference*, 16-18 November, Melbourne, Geoscience Australia, Record 2011/43,77-80.

GSWA, 1990, *Geology and Mineral Resources of Western Australia: Western Australia Geological Survey,* Memoir 3, 827p.

Hickman, A.H., 2012, Wodgina, Western Australia Sheet 2655 (2nd edition), Geological Survey of Western Australia Series Map.

Hocking, R. M., van de Graaff, W. J. E., Blockley, J. G.and Butcher, B. P., 1982, Ajana, W.A.: Western Australia Geological Survey, 1:250 000 Geological Series Explanatory Notes, 27p.

Hot Dry Rocks, 2008, Geothermal energy potential in selected areas of Western Australia (Perth Basin), *a report prepared for the Department of Industry and Resources, Western Australia. Geological Survey of Western Australia, Statutory Petroleum Exploration Report*, G31888 A2, unpublished.

Iasky, R. P., 1993, A structural study of the southern Perth Basin, Western Australia: *Western Australia Geological Survey, Report* 31, 1993.

Iasky, R.P., 1990, A Geophysical Synthesis of the Southern Perth Basin, Western Australia, *M.Sc. Thesis*, School of Physical Sciences, Department of Exploration Geophysics, Curtin University of Technology, 290p.

Jacobs, J.A., Russell, R.D., and Tuzo Wilson, J., 1959, Physics and Geology, McGraw Hill Book Company, Inc., 424p.

Jaeger, J.C., 1970, Heat flow and radioactivity in Australia, *Earth and Planetary Science Letters*, 8, 285-292.

Janssen, D. P., Collins, A. S., and Fitzsimons, I. C. W.,2003,Structure and tectonics of the Leeuwin Complex and Darling Fault Zone, southern Pinjarra Orogen, Western Australia — a field guide: Western Australia Geological Survey, Record 2003/15, 33p.

Jessop, A.M., 1990, *Thermal Geophysics*, Developments in Solid Earth Geophysics, Elsevier, Amsterdam, 306p.

Joshua, E.O., Ehinola, O.A., Akpanowo, M.A., and Oyebanjo, O.A., 2008, Radiogenic heat production in crustal rock samples of Southeastern Nigeria, European Journal of Scientific Research (EuroJournals Publishing Inc.), 23, 305-316

Kappelmeyer, O. and Haenel, R., 1974, *Geothermics with special reference to Application*, Bebrüder Borntraeger.

Lennard, E.L., 1991, *Yallingup Sheet 1930 IV and part Sheet 18301,* Geological Survey of Western Australia.1:50 000 Environmental Geology Series.

Lockwood, A.M., 2004, Western Ausatralia isostatic residual gravity anomaly and depth to basement model, Western Australia Geological Survey, Record 2004/14, 31p.

Mernagh, T.P. and Miezitis, Y., 2008, A Review of the geochemical processes controlling the distribution of thorium in the Earth's crust and Australia's thorium resources.Geoscience Australia Record 2008/05, 48p.

Middleton, M.F. and Stevens, M. K., 2013, *Heat generation in the Vasse Shelf region: implications for geothermal exploration,* Petroleum in Western Australia (PWA), September 2013, 26-32.

Middleton, M.F., 1993, A transient method of measuring thermal properties of rocks, *Geophysics*, 58, 357-365.

Middleton, M.F., 2010, The Gillingarra temperature anomaly, northern Perth Basin, Australia, *Expanded Abstracts, Proceedings of SEG International Exposition and 80th Annual Meeting, Denver, Colorado*, October 2010.

Middleton, M.F., 2011, Temperatures within outcropping, radiogenic granites: Implications for geothermal energy in Western Australia, in: Middleton,

M. and Gessner, K. (eds.) *West Australian Geothermal Energy Symposium Abstracts*, 21 & 22 March 2011, v.1, 20 (ISBN 978-1-74052-230-4).

Middleton, M.F., 2012, Potential for geothermal energy in the Pilbara region from hot dry rocks (HDR) geothermal systems, Petroleum in Western Australia (PWA), September 2012, 28-31.

Middleton, M.F., 2013, Radiogenic heat generation in the Darling Range, Western Australia, *Exploration Geophysics*, 44, 206-214.

Middleton, M.F., Wilde, S.A., Evans, B.J., Long, A., Dentith, M. and Morawa, 1995, Deep seismic reflection traverse over the Darling Fault Zone, Western Australia, *Australian Journal of Earth Science*, 42, 83-93.

Minty, B., Franklin, R., Milligan, P., Richardson, L.M. and Wilford, J., 2009, The radiometric map of Australia, *Exploration Geophysics*, 40, 325-333.

Mory, A.J. & Haines, P.W., 2013, A Paleozoic Perspective of Western Australia, in Keep, M. & Moss, S.J. (Eds), 2013, *The Sedimentary Basins of Western Australia IV*: Proceedings of the Petroleum Exploration Society of Australia Symposium, Perth, WA, 2013.

Mory, A.J., and Iasky, R.P., 1996, Stratigraphy and structure of the onshore northern Perth Basin, Western Australia: Western Australia Geological Survey, Report 46.

Myers, J.S., 1990a, Geological Evolution, Western Gneiss Terrane, in *Geology and Mineral Resources of Western Australia*, Western Australia Geological Survey, Memoir *3,* 13-32.

Myers, J.S., 1990b, Geological Evolution, Precambrian, in *Geology and Mineral Resources of Western Australia*, Western Australia Geological Survey, Memoir ***3,*** 737-749.

Perkins, D., 1994, *The Birdsville Geothermal Electricity Project. Development of a commercial demonstration facility*, Internal Report, Queensland Electricity Commission(QEC), Brisbane, Qld. 11p.

Playford, P. E., Cockbain, A. E., and Low, G. H., 1976,Geology of the Perth Basin, Western Australia: *Western Australia Geological Survey, Bulletin* 124.

RMA, 2010, Radiometric database for Australia, Geoscience Australia, 2010.

RSI, 2009, Radiation Solutions Inc., www.radiationsolutions.ca, Nov. 2009, RS-125 Super Spec, RS 230 BGO Super Spec, Portable Radiation Detector, Version 2.0 – Nov. 2009.

Sanders, A. J., and McGuiness, S. A., 2000, Geochemical mapping of the Ajana 1:250 000 map sheet: Western Australia Geological Survey, 1:250 000 Regolith Geochemistry Series Explanatory Notes, 55p.

Sass, J.H., Jaeger, J.C. and Munroe, R.J., 1976, Heat flow and near-surface radioactivity in the Australian continental crust, *United States Department of the Interior, Geological Survey, Open-File Report* 76-250.

Schön, J.H., 1996, *Physical properties of Rocks: Fundamentals and principles of petrophysics*, Pergamon.

Simpson, E.S., 1928, Famous Mineral Localities: Wodgina, North West Australia, American Mineralogist, V. 13, pages 457-468.

Stolz, N. 2010, New precompetitive data for uranium and geothermal energy exploration in Australia, *Expanded Abstracts, Proceedings of SEG International Exposition and 80th Annual Meeting, Denver, Colorado*, October 2010.

The Radiometric Map of Australia Dataset (radmap10). *Geoscience Australia,* 2010.

Turner, F.J., 1968, *Metamorphic Petrology – Mineralogical and field aspects,* McGraw-Hill Book Company, 403p.

Van Krandendonk, M.J., 2011, personal communication concerning the geochemical database of the Geological Survey of Western Australia.

Van Krandendonk, M.J., Smithies, R. H., Hickman, A.H. and Champion, D.C., (2007), Review: secular tectonic evolution of Archean continental crust: interplay between horizontal and vertical processes in the formation of the Pilbara Craton, Australia, *Terra Nova*, 19, 1-28.

Welch, P., and Boyle, P., 2009, New turbines to enable efficient geothermal power plants, *Geothermal Research Council (GRC) Transactions*, 33, 765-772.

Welch, P., Boyle, P., Sells, M., and Murillo, I., 2010, Performance of new turbines for geothermal power plants, *Geothermal Research Council (GRC) Transactions*, 34, 1091-1096.

Wharton, P.H., 1982, The geology and hydrogeology of the Quindalup boreline, southern Perth Basin, Western Australia, Geological Survey of West Australia, Record 1982/2.

Wichleder, M.S., Fourest, B. And Borhourt, P.K., 2006, Thorium, In: Moress, L.R., Edelstein, N.M. and Fuger, J., The Chemistry of the Actinide and Transactinide Elements (3rd Ed.), Springer, ISBN 1-4020-3555-1.

Wilde, S.A. and Low, G.H, 1978a, Perth, Western Australia, Sheet SH50-14 International Index, 1:250 000 Geological Series Explanatory Notes, *Geological Survey of Western Australia,* ISBN 0 7244 7546.

Wilde, S.A. and Low, G.H, 1978b, Pinjarra, Western Australia, Sheet SH50-2 International Index, 1:250 000 Geological Series Explanatory Notes, *Geological Survey of Western Australia,* ISBN 0 7244 8238 5.

Wilde, S.A., Middleton, M.F. and Evans, B.J., 1996, Terrane accretion in the southwestern Yilgarn Craton: evidence from a deep seismic crustal profile, *Precambrian Research*, 78, 179-196.

In: Thorium
Editor: Renae Jamison
ISBN: 978-1-63321-309-8

Chapter 7

THERMOPHYSICAL AND THERMODYNAMIC PROPERTIES OF OXYGEN-CONTAINING COMPOUNDS OF THORIUM

Alexander V. Knyazev* and Marina E. Manyakina
Lobachevsky State University of Nizhni Novgorod, Russia

ABSTRACT

This chapter presents the chemical thermodynamic data set for thorium species. Analysis of the majority of calorimetric data on thorium oxide compounds is made. Thermal expansion is one of the thermophysical characteristics of the matter. It gives an information about materials behaviour in a wide range of temperature.

Keywords: Thorium; thermodynamic functions; heat capacity; high-temperature X-ray diffraction; thermal expansion coefficient

INTRODUCTION

The modelling of the behaviour of hazardous materials under environmental conditions is among the most important applications of natural

* Corresponding Author email address: knyazevav@gmail.com.

and technical sciences for the protection of the environment. In order to assess, for example, the safety of a waste deposit, it is essential to be able to predict the eventual dispersion of its hazardous components in the environment (geosphere, biosphere). For hazardous materials stored in the ground or in geological formations, the most probable transport medium is the aqueous phase. An important requirement for predicting the pathways and rates of aqueous transport of potential contaminants is therefore the quantitative prediction of the reactions that are likely to occur between hazardous waste dissolved or suspended in ground water, and the surrounding rock material, in order to estimate the quantities of waste that can be transported in the aqueous phase. It is thus essential to know the relative stabilities of the compounds and complexes that may form under the relevant conditions. This information is often provided by speciation calculations using chemical thermodynamic data. The local conditions, such as ground water and rock composition or temperature, may not be constant along the migration paths of hazardous materials, and fundamental thermodynamic data are the indispensable basis for dynamic modeling of the chemical behaviour of hazardous waste components.

In the field of radioactive waste management, the hazardous material consists to a large extent of actinides and fission products from nuclear reactors, in addition to lesser amounts from other sources such as waste from medicine, industry and research facilities. The scientific literature on thermodynamic data, mainly on equilibrium constants and redox potentials in aqueous solution, has been contradictory in a number of cases, especially in the actinide chemistry. A critical and comprehensive review of the available literature is necessary in order to establish a reliable thermochemical data base that fulfils the requirements of a proper modelling of the behaviour of the actinide and fission products in the environment.

At the present time, much attention is paid to the possibility of using thorium in various schemes and scenarios of nuclear energy. In the absorption of slow neutrons thorium nuclear reaction occurs next: $^{232}Th\ (n, \gamma) \rightarrow {}^{233}Th \rightarrow {}^{233}Pa \rightarrow {}^{233}U$. The resulting ^{233}U, as now used ^{235}U, capable of sustaining a chain reaction. Thorium fuel cycle has several advantages over the applicable uranium and uranium-plutonium cycle. Thorium content in the crust of uranium exceeds four times. Almost all natural thorium consists of the target isotope ^{232}Th, while the content of ^{235}U is only 0.7%. Accumulation of long-lived minor actinides reduced compared to uranium-plutonium fuel cycle. This environmentally positive factor has been recently used as the main argument in favor of thorium. At the moment, the biggest problem for the introduction of the thorium fuel cycle for commercial power production is the lack of infrastructure related to the fabrication of fuel and waste disposal technology thorium cycle. Decision radiochemical problems are impossible without fundamental information about the oxygen-containing compounds of thorium. The knowing of dependences "composition-structure-properties" is of great importance.

In this chapter, the thermodynamic functions and the coefficients of thermal expansion are described.

THERMODYNAMIC PROPERTIES OF OXYGEN-CONTAINING COMPOUNDS OF THORIUM

The two Gmelin volumes on minerals (R. Ditz, B. Sarbas, P. Schubert, 1990), (F.H. Bartel, F.J. Dahlkamp, R. Ditz, et al, 1991) provide a good review of thorium minerals (figure 1).

In the earth's crust, thorium is about three times as abundant as uranium. In minerals, thorium is frequently associated with other quadrivalent species (Zr, Hf, Ce, and U), but also with tervalent lanthanides. Table 1 gives a list of common thorium containing minerals, extracted from the Mineralogical Database website1.

Figure 1. Mineral forms of thorium oxigen-containing compounds.

Table 1. Thorium minerals

Mineral name	Chemical formula	Th content (wt%) in a typical sample
Althupite	$ThAl(UO_2)_7(PO_4)_4(OH)_5\cdot 15H_2O$	8.04
Arapovite	$(U,Th)(Ca,Na)_2(K1_{-x[]x})Si_8O_{20}$, x~0.5	7.07
Brabantite	$CaTh(PO_4)_2$	50.22
Britholite-(Ce)	$(Ce,Ca,Th,La,Nd)_5(SiO_4,PO_4)_3(OH,F)$	18.25
Brockite	$(Ca,Th,Ce)(PO_4)\cdot H_2O$	31.55
Calciobetafite	$Ca_2(Ti,Nb)_2(O,OH)_7$	5.88
Cerianite-(Ce)	$(Ce,Th)O_2$	29.73
Cervandonite-(Ce)	$(Ce,Nd,La)(Fe^{+3},Fe^{+2},Ti^{+4},Al)_3SiAs(Si,As)O_{13}$	3.66
Cheralite-(Ce)	$(Ce,Ca,Th)(P,Si)O_4$	29.02
Chevkinite-(Ce)	$(Ce,La,Ca,Th)_4(Fe^{+2},Mg)_2(Ti,Fe^{+3})_3Si_4O_{22}$	1.93
Ciprianiite	$Ca_4[(Th,U,Ln]_2(Al,Va)_2[Si_4B_4O_{22}](OH,F)_2$	14.26
Coutinhoite	$ThxBa_{1-2x}(H_2O)_y(UO_2)_2Si_5O_{13}\cdot H2O$	3.29
Dissakisite-(La)	$(Ca,Fe,Th)(REE,Ca)(Al,Cr,Ti)_2(Mg,Fe,Al)Si_3O_{12}(OH,F)$ with La>Ce	3.98
Ekanite	ThCa2Si8O20	27.08
Eylettersite	$(Th,Pb)_{(1-x)}\ Al_3(PO_4,SiO_4)_2(OH)_6(?)$	19.29
Fersmite	$(Ca,Ce,Na)(Nb,Ta,Ti)_2(O,OH,F)$	2.97
Grayite	$(Th,Pb,Ca)PO_4\cdot H_2O$	43.73
Huttonite	$ThSiO_4$	71.59
IMA2004-001	$[(REE^{+}Y),U,Th,Ca,Fe](Nb,Ta,Ti)O_4$	6.83
Ilimaussite-(Ce)	$(Ba,Na)_{10}K_3Na_{4.5}Ce_5(Nb,Ti)_6[Si_{12}O_{36}][Si_9O_{18}(O,OH)_{24}]O_6$	1.04
Iraqite-(La)	$K(La,Ce,Th)_2(Ca,Na)_4(Si,Al)_{16}O_{40}$	9.94
Karnasurtite-(Ce)	$(Ce,La,Th)(Ti,Nb)(Al,Fe^{+3})(Si,P)_2O_7(OH)_4\cdot 3H_2O(?)$	6.18
Kivuite	$(Th,Ca,Pb)H_2(UO_2)_4(PO_4)_2(OH)_8\cdot 7H_2O$	8.16
Kuannersuite-(Ce)	$Ba_6Na_2REE_2(PO_4)_6FCl$	0.27
Kukharenkoite-(La)	$Ba_2(La,Th,Ce)(CO_3)_3F$	10.87
Melanocerite-(Ce)	$(Ce,Th,Ca)_5(Si,B)_3O_{12}(OH,F)\cdot nH_2O(?)$	5.14
Monazite-(Ce)	$(Ce,La,Nd,Th)PO_4$	4.83
Monazite-(La)	$(La,Ce,Nd)PO_4$	4.84
Monazite-(Nd)	$(Nd,Ce,La)(P,Si)O_4$	4.82
Monazite-(Sm)	$SmPO_4$	15.73
Mottanaite-(Ce)	$Ca_4(Ce,Ca)_2AlBe_2[Si_4B_4O_{22}]O_2$	2.30
Okanoganite-(Y)	$(Na,Ca)_3(Y,Ce)_{12}Si_6B_2O_{27}F_{14}$	1.66
Orthochevkinite	$(Ce,La,Ca,Na,Th)_4(Fe^{+2},Mg)_2(Ti,Fe^{+3})_3Si_4O_{22}$	7.93
Perrierite-(Ce)	$(Ce,La,Ca)_4(Fe^{+2},Mg)_2(Ti,Fe^{+3})_3Si_4O_{22}$	0.97
Polyakovite-(Ce)	$(Ce,La,Nd,Pr,Ca)_4(Mg,Fe^{++})(Cr,Fe^{+++})2(Ti,Nb)_2Si_4O_{22}$	1.85
Polycrase-(Y)	$(Y,Ca,Ce,U,Th)(Ti,Nb,Ta)_2O_6$	6.22
Saryarkite-(Y)	$Ca(Y,Th)Al_5(SiO_4)_2(PO_4,SO_4)_2(OH)_7\cdot 6H_2O$	7.53
Steacyite	$K_{1-x}(Ca,Na)_2ThSi_8O_{20}$(x=0.2 to 0.4)	25.29

Mineral name	Chemical formula	Th content (wt%) in a typical sample
Steenstrupine-(Ce)	$Na_{14}Ce_6Mn^{+2}Mn^{+3}Fe_2^{+2}(Zr,Th)(Si_6O_{18})_2(PO_4)_7 \cdot 3H_2O$	1.85
Thorbastnasite	$Th(Ca,Ce)(CO_3)_2F_2 \cdot 3H_2O$	45.57
Thorianite	ThO_2	87.88
Thorite	$ThSiO_4$	71.59
Thornasite	$Na_{12}Th_3[Si_8O_{19}]_4 \cdot 18H_2O$	20.41
Thorogummite	$Th(SiO_4)_{1-x}(OH)_{4x}$	72.13
Thorosteenstrupine	$(Ca,Th,Mn)_3Si_4O_{11}F \cdot 6H_2O$	29.29
Thorutite	$(Th,U,Ca)Ti_2(O,OH)_6$	23.75
Tritomite-(Ce)	$(Ce,La,Ca,Y,Th)_5(Si,B)_3(O,OH,F)_{13}$	4.42
Tuliokite	$BaNa_6Th(CO_3)_6 \cdot 6H_2O$	23.79
Turkestanite	$Th(Ca,Na)_2$(K1-x,Vax)Si8O20·nH2O	21.11
Umbozerite	$Na_3Sr_4ThSi_8(O,OH)_{24}$	18.33
Uranopolycrase	$(U,Y)(Ti,Nb,Ta)_2O_6$	5.58
Vicanite-(Ce)	$(Ca,Ce,La,Th)_{15}As^{+5}(As^{+3}{}_{0.5},Na_{0.5})Fe^{+3}Si_6BO_{40}F_7$	9.08
Yttrialite-(Y)	$(Y,Th)_2Si_2O_7$	27.79
Yttrocrasite-(Y)	$(Y,Th,Ca,U)(Ti,Fe^{+3})_2(O,OH)_6$	7.79
Zirconolite-3O	$(Ca,Fe,Y,Th)Fe(Ti,Nb)_3Zr_2O_7$	3.61
Zirkelite	$(Ca,Th,Ce)Zr(Ti,Nb)_2O_7$	6.06

In nature, ThO_2 is found mostly as solid solutions with UO_2. Conventionally, the name thorianite is used when the molar content of ThO_2 is above 75%; uranothorianite is used for such solid solutions containing 25 - 75% ThO_2; thorian uraninite is employed when the ThO_2 content is 15 - 25%; uraninite is used when the UO_2 content is above 85%.

In addition, it should be noted that, although the generic formula of someminerals listed in Table 1 does not mention thorium, this element is often present in such minerals by substitution for quadrivalent or tervalent species, in percentages that may be in excess of 10%. This is, for instance, the case for Sm-monazites. Of course, in such substitutions charge balance must be maintained and the replacement of a tervalent ion by thorium must be accompanied by an equivalent increased amount of a lowervalent cation (often Ca^{2+}) in the structure.

The thorium content given in the third column of the list is that found for a typical sample.

Table 2 contains the thermodynamic data of the thorium compounds and species. Of greatest interest are thorium oxide compounds and ionic forms.

Table 2. Thermodynamics values of of thorium oxigen-containing compounds

Compound	Condition	$\Delta_f H^\circ_{298,15}$	$\Delta_f G^\circ_{298,15}$	$S^\circ_{298,15}$	$C^\circ p_{298,15}$	Phase transitions		Reference
		kJ/mol		J/(mol·K)		Type	T° K	
Th	cr II', cubic (α)	-	-	-	-	crII' → g	0	(A.J. Darnell, et al., 1960; K.A. Gschneidner, et al., 1962; D.C. Wallace, 1960)
	cr II', cubic (α)	-	-	-	-	crII' → crII'	1,37	
	cr II', cubic(α)	-	-	-	-	crII' → crII'	0,9±0,06	
	cr II', cubic (α)	-	-	-	-	crII' → crII'	0,7±0,05	
	cr II', cubic (α)	-	-	-	-	crII' → crII'	0,7±0,05	
	cr II, cubic (α)	0	0	53.387±0.148	27.321±0.125	crII → g	298,15	
	cr II, cubic (α)	-	-	-	-	cr II → crI	1638	
	cr I, cubic (β)	-	-	-	-	crI → l		
	cr I, cubic (β)	-	-	-	-	crI → g	2023	
	l	-	-	-	-	l → g		
	l	-	-	-	-	l → g	5110±150	
	g	594.128±6.276	553.380	190.058±0.029	20.790±0.004	-	-	
Th^+	g	1323.863	1277.425	188.275±0.029	21.304±0.004	-	-	(J. Sugar, 1973)
Th^{+2}	g	-	-	-	-	-	-	(W. Finkelnburg, W. Humbach, 1955)
Th^{+3}	g	-	-	-	-	-	-	(P.F.A. Klinkenberg, 1950)
Th^{+4}	g	-	-	-	-	-	-	(H.W. Dodgen, G.K. Rolefson, 1949)
	solution in H_2O	-782.408±8.368	-719.572±8.786	-418.4±16.736	-	-	-	
Th_2	g	-	-	-	-	-	-	(K.A. Gingerich, 1969)
ThO	cr, cubic	-	-	62.76±8.368	-	-	-	(E.F. Westrum, F. Gronvold, 1962)
	g	-26.882	-51.940	239.952±0.083	31.254±0.041	-	-	
ThO_2	cr, cubic (thorianite)	-1226.75±1.67	-1169.147	65.228±0.083	61.755±0.627	cr → g	0	(R.J. Ackermann, et al., 1963; M. Hoch, H.L. Johnston, 1961; D.L. Hildenbrand, E.
	cr, cubic (thorianite)	-	-	-	-	cr → l	3623	
	g	-438.499	-447.152	287.440±3.347	47.279±0.418	-	-	

Compound	Condition	$\Delta_f H^\circ_{298,15}$	$\Delta_f G^\circ_{298,15}$	$S^\circ_{298,15}$	$C^\circ p_{298,15}$	Phase transitions		Reference
		kJ/mol		J/(mol·K)		Type	T° K	
								Murad, 1974)
$ThOH^{+3}$	solution in H_2O	-	-940.144±8.786	-	-	-	-	(D.L. Hildenbrand, E. Murad, 1974; K.A. Kraus, R.W. Holmberg, 1954; . E. Matijevic, et al., 1960)
$Th(OH)_2^{+2}$	solution in H_2O	-	-1163.988±8.786	-	-	-	-	
$Th(OH)_4$	solution in H_2O	-1702.55±8.37	-1348.846±8.786	-461.829±16.736		-	-	(V.P. Glushko, 1965-1982)
$Th_2(OH)_2^{+6}$	solution in H_2O	-	-1881.963±17.572	-	-	-	-	(P.R. Danesi, et al., 1968; K.A. Kraus, R.W. Holmberg, 1954)
ThOF	g	-550±12	-566.186±12.366	311±1	49±5	-	-	(Rand, M., et al., 2008)
$ThOF_2$	cr	-1663.8±7.8	-1589.171±7.948	107.7±5.1	-	-	-	
$ThClO_3^{+3}$	solution in H_2O	-	-728.852	-	-	-	-	(R.A. Day, R.W. Stoughton, 1950)
$ThOCl_2$	cr, orthorhombic	-1237.21±6.28	-	-	-	-	-	(K.W. Bagnall, et al., 1968)
$Th(ClO_4)_4$	solution in H_2O	-1276.789±8.786	-733.162±9.623	316.310±18.828	-	-	-	(V.P. Glushko, 1965-1982)
$Th(ClO_4)_4 \cdot 4H_2O$	cr, orthorhombic	-	-	-	-	cr → l	475	(P.R. Murthy, C.C. Patel, 1963)
$ThCl_2OH \cdot H_2O$	cr	-1600.38	-	-	-	-	-	(E. Chauvenet, 1911)
$Th(BrO_3)^{+3}$	solution in H_2O	-	-729.689	-	-	-	-	(R.A. Day, R.W. Stoughton, 1950)

Table 2. (Continued)

Compound	Condition	$\Delta_f H^\circ_{298,15}$	$\Delta_f G^\circ_{298,15}$	$S^\circ_{298,15}$	$C^\circ p_{298,15}$	Phase transitions		Reference
		kJ/mol		J/(mol·K)		Type	T° K	
$Th(BrO_3)_2^{+2}$	solution in H_2O	-	-733.873	-	-	-	-	(R.A. Day, R.W. Stoughton, 1950)
$ThOBr_2$	cr	-1146.416	-	-	-	-	-	(E. Chauvenet, 1911)
ThB_2O_5	cr, monoclinic					cr → l	1763	(Y. Baskin, et al., 1961)
$ThIO_3^{+3}$	solution in H_2O	-	-884.916	-	-	-	-	(R.A. Day, R.W. Stoughton, 1950; R.W. Stoughton, et al., 1961)
$ThOI_2$	cr	-990.352±4.184	-	-	-	-	-	(E. Chauvenet, 1911; D.E. Scaife, et al., 1965)
$ThOI_2 \cdot 3{,}5H_2O$	cr	-2038.863	-	-	-	-	-	(E. Chauvenet, 1911)
$Th(IO_3)_2^{+2}$	solution in H_2O	-	-1042.652	-	-	-	-	(R.A. Day, R.W. Stoughton, 1950; R.W. Stoughton, et al., 1961)
$Th(IO_3)_3^{+}$	solution in H_2O.	-	-1201.226	-	-	-	-	
$Th(IO_3)_6^{-2}$	solution in H_2O	-	-1647.659	-	-	-	-	
$Th(IO_3)_8^{-4}$	solution in H_2O	-	-1931.334	-	-	-	-	
$ThOH(IO_3)^{+2}$	solution in H_2O	-	-1106.249	-	-	-	-	
$ThOH(IO_3)_2^{+}$	solution in H_2O	-	-1268.170	-	-	-	-	
$ThSO_4^{+2}$	solution in H_2O	-	-1504.984± 8.786	-	-	-	-	(A.J. Zielen, 1959)
$Th(SO_3)_2$	cr	-	-	-	-	cr → l	1283	(N.N. Mironov, 1959)
$Th(SO_4)_2$	cr	-	-	145.603±2.092	173.217± 2.092	-	-	. (S.W. Mayer, et al., 1960)
	solution in H_2O	-2604.37±8.37	-2210.875±8.	-382.417± 16.736	-	-	-	
	solution in H_2O	-	-2266.054± 8.786	-	-	-	-	
$Th(SO_4)_2 \cdot 4H_2O$	cr, orthorhombic	-	-3209.964± 8.786	-	-	-	-	(J. Koppel, 1910)

Compound	Condition	$\Delta_f H^\circ_{298,15}$	$\Delta_f G^\circ_{298,15}$	$S^\circ_{298,15}$	$C^\circ p_{298,15}$	Phase transitions		Reference
		kJ/mol		J/(mol·K)		Type	T° K	
$Th(SO_4)_2 \cdot 8H_2O$	cr1, monoclinic	-	-4170.192± 8.786	-	-	crI → crII	298	(R.R. Sood, R.A. Stage, 1966)
$Th(SO_4)_2 \cdot 9H_2O$	cr	-	-4408.471± 8.786	-	-	-	-	(V.P. Glushko, 1965-1982)
$Th(SO_4)_3^{-2}$	solution in H_2O	-	-3017.919± 8.786	-	-	-	-	(K.A. Allen, W.J. McDowell, 1963)
$Th(S_2O_3)_2$	cr	-	-	-	-	cr → l	1373	(N.N. Mironov, 1959)
ThOSe	cr, tetragonal	-	-	-	-	cr → l	2473	R.W.M. D'Eye, et al., 1952)
$Th(SeO_3)_2$	cr	-	-1560.213	-	-	-	-	(E.I. Krylov, V.G. Chukhlantsev, 1957)
$ThNO_3^{+3}$	solution in H_2O	-	-843.912±8.786	-	-	-	-	(J. Danon, 1960; R.A. Day, R.W. Stoughton, 1950)
$Th(NO_3)_2^{+2}$	solution in H_2O	-	-966.922±8.786	-	-	-	-	
$Th(NO_3)_3^{+}$	solution in H_2O	-	-1081.564± 8.786	-	-	-	-	
$Th(NO_3)_4$	cr	-1461.471	-1064.371	334.72±20.92	-	-	-	(J. Danon, 1960; R.A. Day, R.W. Stoughton, 1950; J.R Ferraro, et al., 1956)
	solution in H_2O	-1612.01±8.79	-1165.921± 9.204	170.422± 19.246	-	-	-	
	solution in H_2O		-1195.368± 8.786	-	-	-	-	
$Th(NO_3)_4 \cdot 4H_2O$	cr	-2718.763	-2093.640	502.08± 8.368	-	cr → l + cr_1	426	(W.L. Marshall, et al., 1951)
$Th(NO_3)_4 \cdot 5H_2O$	cr, orthorhombic	-3021.68±8.37	-2339.341	543.208±0.627	480.825± 0.418	-	-	(J.A.R. Cheda, et al., 1976; J.R Ferraro, et al., 1956)
$Th(NO_3)_4 \cdot 6H_2O$	cr	-	-	-	-	cr → l + cr_1	384,45	(W.L. Marshall, et al., 1951)

Table 2. (Continued)

Compound	Condition	$\Delta_f H^{\circ}_{298,15}$	$\Delta_f G^{\circ}_{298,15}$	$S^{\circ}_{298,15}$	$C^{\circ}p_{298,15}$	Phase transitions		Reference
		kJ/mol		J/(mol·K)		Type	T° K	
Th_2N_2O	cr, hexagonal	-1345.574	-1255.894	99.997	-	-	-	(T. Kusakabe, S. Imoto, 1972)
$ThNO_3SO_4^+$	solution in H_2O	-	-1627.157± 8.786	-	-	-	-	(E.P. Mayorova, V.V. Fomin 1958; Mayorova E.P., Fomin V.V. 1959)
$Th(NO_3)_2SO_4$	solution in H_2O	-	-1735.104± 8.786	-	-	-	-	
$Th(NO_3)_3SO_4^-$	solution in H_2O	-	-1848.491± 8.786	-	-	-	-	
ThP_2O_7	cr IV, cubic (α)	-	-	-	-	crIV→ crIII	371	(G. Busch, et al., 1965; N.N. Mironov, 1959)
	cr III, cubic (α)	-	-	-	-	crIII→ crII	1083	
	cr II, orthorhombic (β)	-	-	-	-	crII→ crI	1563	
	cr I, cubic	-	-	-	-	crI→ l	3263	
$Th_4(PO_4)_4P_2O_7$	cr	-	-	-	569±15	-	-	(Rand, M., et al., 2008)
$ThH_2PO_4^{3+}$	solution in H_2O	-	-1873.843± 5.822	-	-	-	-	
$ThH_3PO_4^{4+}$	solution in H_2O	-	-1864.938± 5.804	-	-	-	-	
$Th(H_2PO_4)_2^{2+}$	solution in H_2O	-	-3038.907± 6.430	-	-	-	-	
$Th(H_3PO_4)(H_2PO_4)^{3+}$	solution in H_2O	-	-3034.455± 6.430	-		-	-	
$Th_3(AsO_3)_4$	cr	-	-	-	-	cr → l	1046	(V.P. Glushko, 1965-1982)
$Th_3(AsO_4)_4$	cr I, orthorhombic (α)	-	-	-	-	crI→ l	1052	
$ThC_2O_4^{+2}$	solution in H_2O	-	-1413.773	-	-	-	-	(H. Bilinski, N. Ingri, 1967)
$Th(C_2O_4)_2$	solution in H_2O	-2431.91±8.79	-2057.331±	-359.824±	-	-	-	(H. Bilinski, N. Ingri,

Compound	Condition	$\Delta_f H^\circ_{298,15}$	$\Delta_f G^\circ_{298,15}$	$S^\circ_{298,15}$	$C^\circ p_{298,15}$	Phase transitions		Reference
		kJ/mol		J/(mol·K)		Type	T° K	
			10.46	23.848				1967;
	solution in H_2O	-	-2104.970	-	-	-	-	R. Bressat, et al., 1963)
$Th(C_2O_4)_2\cdot 6H_2O$	cr	-	-3594.056	-	-	-	-	(R. Bressat, et al., 1963)
$Th(C_2O_4)_3^{-2}$	solution in H_2O	-	-2800.351	-	-	-	-	(H. Bilinski, N. Ingri, 1967)
$Th(C_2O_4)_4^{-4}$	solution in H_2O	-	-3525.856	-	-	-	-	
$ThHCOO^{+3}$	solution in H_2O	-	-1095.789± 8.786	-	-	-	-	(R. Portanova, et al., 1975; R. Portanova, et al., 1972; G.M. Sergeev et al., 1970; P.H. Tedesco, M.C. Anon, 1972)
$Th(HCOO)_2^{+2}$	solution in H_2O	-	-1464.4± 8.786	-	-	-	-	
$ThCH_3COO^{+3}$	solution in H_2O	-1259.38± 8.37	-1117.964± 8.768	-204.413± 41.003	-	-	-	
$Th(HCOO)_3^{+}$	solution in H_2O	-	-1828.408± 8.786	-	-	-	-	
$Th(CH_3COO)_2^{+2}$	solution in H_2O	-1743.05± 8.37	-1510.424± 8.786	-32.777± 41.003	-	-	-	
$Th(CH_3COO)_3^{+}$	solution in H_2O	-2217.938± 8.368	-1894.515± 8.786	140.260± 41.003	-	-	-	
$Th(CH_3COO)_4$	solution in H_2O	-2700.772± 8.368	-2273.167± 8.786	268.391± 41.003	-	-	-	(R. Portanova, et al., 1975)
$Th(CH_3COO)_5^{-}$	solution in H_2O	-3185.279± 8.368	-2646.798± 8.786	376.070± 41.003	-	-	-	

Table 2. (Continued)

Compound	Condition	$\Delta_f H°_{298,15}$	$\Delta_f G°_{298,15}$	$S°_{298,15}$	$C°p_{298,15}$	Phase transitions		Reference
		kJ/mol		J/(mol·K)		Type	$T°$ K	
$ThSiO_4$	cr II, tetragonal (torite)	-	-2049.323± 5.020	-	-	crII→ crI	1498	(R.D. Schuiling, 1976)
	cr I, monoclinic (huttonite)	-	-	-	-	crI→ l + cr_1	2248	
$ThTi_2O_6$	cr II, monoclinic (torutite)	-3095.0±4.3	-	-	-	crII→ crI	1573	(A. Harari, 1967; A. Revcolevschi,1970)
$ThGeO_4$	cr I	-	-	-	-	crI→ l + cr_1	1693	(L.A. Harris, C.B. Finch, 1972)
$Th(CrO_4)^{+2}$	solution in H_2O	-366,6	-358,2	57,051	-	-	-	(M.J. Burkhart, R.C. Thompson, 1972)
$ThO_2 \cdot 2MoO_3$	cr II, orthorhombic (β)	-	-	-	-	crII→ crI	1173	(V.P. Glushko, 1965-1982)
$ThO_2 \cdot 2Nb_2O_5$	cr, tetragonal	-	-	-	-	cr → l	1661	(C. Keller, 1965)
$2ThO_2 \cdot Nb_2O_5$	cr II	-	-	-	-	crII→ crI	1583	
	cr I	-	-	-	-	crI→ l	1614	
$Na_6Th(CO_3)_5 \cdot 12 H_2O$	cr	-	-8002.562± 7.301	-	-	-	-	(Rand, M., et al., 2008)
$LiTh_2(AsO_4)_3$	cr	-	-	-	-	cr→l + crl	1533	(W. Freundlich, et al., 1974)
$Li_2Th(AsO_4)_2$	cr	-	-	-	-	cr→l + crl	1433	
$NaTh_2(AsO_4)_3$	cr	-	-	-	-	cr→l + crl	1623	
$Na_2Th(AsO_4)_2$	cr	-	-	-	-	cr → l	1423	
$CaThNb_2O_8$	cr	-	-	-	-	crII→ crI	1023	(G. Fontenau, L.J. L'Helgoulach, 1977)
$SrThNb_2O_8$	cr	-	-	-	-	crII→ crI	808	
$Li_2Th(MoO_4)_9$	cr	-	-	-	-	cr→l + crl	1283	(J. Thoret, 1971)
$Na_2Th(MoO_4)_3$	cr	-	-	-	-	cr→l + crl	1161	
$Na_4Th(MoO_4)_4$	cr	-	-	-	-	cr→l + crl	2091	
$K_2Th(MoO_4)_3$	cr	-	-	-	-	cr→l + crl	1053	(V.P. Glushko, 1965-

Compound	Condition	$\Delta_f H^\circ_{298,15}$	$\Delta_f G^\circ_{298,15}$	$S^\circ_{298,15}$	$C^\circ p_{298,15}$	Phase transitions		Reference
		kJ/mol		J/(mol·K)		Type	T° K	
$K_4Th(MoO_4)_4$	cr	-	-	-	-	cr→l + crl	978	1982;
$K_8Th(MoO_4)_6$	cr	-	-	-	-	cr→l + crl	1011	J. Thoret, 1971)
$Rb_2Th(MoO_4)_3$	cr	-	-	-	-	cr → l	1018	(V.P. Glushko, 1965-
$Rb_4Th(MoO_4)_4$	cr	-	-	-	-	cr → l	988	1982;
$Rb_8Th(MoO_4)_6$	cr	-	-	-	-	cr → l	988	N.N. Bushuev, V.K. Trunov, 1975)
$Cs_2Th(MoO_4)_3$	cr	-	-	-	-	cr → l	978	(V.P. Glushko, 1965-
$Cs_4Th(MoO_4)_4$	cr	-	-	-	-	cr → l	958	1982)

Thorium as a Simple Substance

Thorium metal has a face centred cubic structure, (space group Fm3m , Cu type) at 298.15 K and undergoes a structural change to a bcc-phase (space group Im3m ,W type) at 1650 K. However, the small amounts of carbon, nitrogen and oxygen, which are difficult to remove from thorium, raise the transformation temperature, whereas additions of niobium, tantalum and zirconium tend to lower the temperature of the phase change. The melting point and enthalpy of fusion of thorium, selecting T_{fus} = 2023 ± 10 K from eight studies and $\Delta_{fus}H$ (Th, 2023 K) = 13.8 ± 0.1 kJ·mol^{-1}. The heat capacity data for Th(cr) are rather unsatisfactory, especially those from 4 to 30 K and a more definitive study with high purity thorium of the values below 300 K is clearly required. In our opinion, the most reliable studies can be represented in the temperature intervals.

- From 0 to 4 K: Observed at the superconducting transition at temperature 1.374 K.
- From 4 to 20 K: data fitted to a cubic equation in T, so that C_p and dC_p/dT are continuous with the values from (J. E.Gordon, H. Montgomery et al., 1966) at 4 K.
- From 20 to 80 K: data of (M. Griffel, R.E. Skochdopole, 1953).
- From 80 to 300 K: data of (J. Nakamura, Y. Takahashi, 1980).

Thermodynamic Data for the Th^{4+} Ion

The standard enthalpy of formation of Th^{4+}, $\Delta_f H^\circ_m$ (Th^{4+}, 298.15 K), is obtained from $\Delta_f H^\circ_m$ ($ThCl_4$, β, 298.15 K), and the extrapolation of the enthalpy of solution of $ThCl_4$ in HCl solutions to infinite dilution, according to reaction: β-$ThCl_4$ ↔ Th^{4+} + 4Cl^-. The enthalpy of solution of β-$ThCl_4$ as a function of the molality (m) of the HCl solutions can be represented a linear extrapolation: $\Delta_{sol}H^\circ$($ThCl_4$, β) = -250.658 + 9.4002 m kJ·mol^{-1}.

yields at zero ionic strength the selected value $\Delta_{sol}H^\circ$($ThCl_4$, β, 298.15 K) = -250.7 ± 1.8 kJ·mol^{-1}, the uncertainty being the 2σ value from the linear regression; hence, using $\Delta_f H^\circ_m$ (Cl^-, 298.15 K) = -167.080 ± 0.100 kJ·mol^{-1} from (J.D. Cox, 1989) and $\Delta_f H^\circ$($ThCl_4$, β, 298.15 K) = -1186.3 ± 1.3 kJ·mol^{-1}, $\Delta_f H^\circ$(Th^{4+}, 298.15 K) = -768.7 ± 2.3 kJ·mol^{-1}.

Thorium Oxide

ThO_2 has a face centred cubic fluorite structure (space group *Fm3m*). Lattice parameters varying from 5.59525 to 5.5997 Å have been reported for ThO_2 (cr), the variation perhaps being due to small differences in stoichiometry and/or the presence of impurities.

The enthalpy of formation is taken from the only modern determination, by Huber *et al.* (Jr. E. J. Huber 1952), who combusted well-characterised Th(cr) whose main impurity was 0.12 wt% oxygen. Their reported value -1226.9 ± 1.5 $kJ\cdot mol^{-1}$ becomes -1226.4 ± 1.5

$kJ\cdot mol^{-1}$ through the use of contemporary atomic masses. This value was adopted by CODATA (J.D. Cox, 1989) with uncertainty limits increased to ±3.5 $kJ\cdot mol^{-1}$ and these are selected by this review: $\Delta_f H°(ThO_2$, cr, 298.15 K) = -1226.4 ± 3.5 $kJ\cdot mol^{-1}$.

The only study of the low temperature heat capacity is made measurements from 10.2 to 305.4 K (D. W. Osborne, Jr. E. F. Westrum, 1953). As expected, no anomaly was observed in the heat capacity. The heat capacity and derived entropy, corrected for modern atomic masses: $C°_p(ThO_2$, cr, 298.15K) = 61.74 ± 0.15 $J\cdot K^{-1}\cdot mol^{-1}$

$S°(ThO_2$, cr, 298.15 K) = 65.23 ± 0.20 $J\cdot K^{-1}\cdot mol^{-1}$ are the selected values, with increased uncertainties, following (J.D. Cox, 1989).The above selections yield: $\Delta_f G°(ThO_2$, cr, 298.15 K) = -1169.0 ± 3.5 $kJ\cdot mol^{-1}$.

Binary Oxides

Thorium Silicate

$ThSiO_4$(cr) exists in two polymeric forms, thorite and huttonite. Thorite is tetragonal with a = 7.1328 Å, c = 6.3188 Å, while huttonite is monoclinic, with a = 6.80 Å, b = 6.96 Å, c = 6.54 Å, β = 104.55°. However, the precise form of the polymorphism is still not clear. Huttonite is certainly the stable form at high temperatures and pressures, and there are two independent studies, which suggest that the equilibrium temperature is 1490 ± 10 K. Measurements of the enthalpy difference of samples of thorite and huttonite between 298.15 and 1774 K gives the enthalpy of the transition of thorite to huttonite, since thorite is transformed into huttonite at the high temperature. The experimental data give a value of $\Delta_{trs}H_m$(298.15 K) = 7.2 ± 7.4 $kJ\cdot mol^{-1}$.

The only selected values are the enthalpies of formation from the study by (L. Mazeina, S. V.Ushakov, 2005): $\Delta_r H°(ThSiO_4$, thorite, 298.15 K) = -2117.5 ± 4.2 kJ·mol^{-1}

$\Delta_r H°(ThSiO_4$, huttonite, 298.15 K) = -2110.8 ± 4.7 kJ·mol^{-1}.

Thorium Titanate (Brannerite)

Thorium frequently substitutes for uranium in the mineral brannerite of generic formula (U,Ca,Ce)(Ti,Fe)$_2$O$_6$. Hence $ThTi_2O_6$(cr) is sometimes referred to as Th-brannerite. Its enthalpy of formation has been measured by high temperature drop calorimetry at 976 K, using a sodium molybdate solvent with the composition $3Na_2O \cdot 4MoO_3$. X-ray powder diffraction of the compounds showed them to be monoclinic, space group *C2/m* with a = 9.8046 ± 0.0008 Å, b = 3.8187 ± 0.0003 Å, c = 7.0229 ± 0.0005 Å, and β = 118.852 ± 0.005°. The value of 19.4 ± 1.6 kJ·mol^{-1} was reported for reaction corresponding to the enthalpy of the formation reaction of brannerite from the oxides, ThO_2(cr) + 2TiO_2(rutile) → $ThTi_2O_6$(cr). Taking their value, it was calculated: $\Delta_f H°_m(ThTi_2O_6$, cr) = -3095.0 ± 4.3 kJ·mol^{-1}.

Thorates

Rb_2ThO_3(cr) is the only known binary oxide in the Rb_2O-ThO_2 system. This compound have a hexagonal structure with a = 3.75 Å, c = 19.7 Å (Rb_2PbO_3 type) and a cubic modification with a = (5.475 ± 0.005) Å (NaCl type) isomorphous with Cs_2ThO_3. The Gibbs energy of the formation reaction of Rb_2ThO_3(cr) from the component oxides is: Rb_2O(l) + ThO_2(cr) → Rb_2ThO_3(cr) $\Delta_r G(T)$ = - 164320 + 2.420T J·mol^{-1}.

Cs_2ThO_3(cr) is the only known binary oxide in the Cs_2O-ThO_2 system and it have a cubic structure with a = 5.704 Å (NaCl type). The Gibbs energy of the formation reaction of Cs_2ThO_3(cr) from the component oxides is: Cs_2O(l) + ThO_2(cr) → Cs_2ThO_3(cr)

$\Delta_r G(T)$ = - 114224 + 25.254T J·mol^{-1}.

The structure of $SrThO_3$(cr) is a monoclinic distortion of the ideal perovskite lattice with a = 6.319 ± 0.004 Å, b = 3.240 ± 0.001 Å and c = 4.928 ± 0.003 Å, β = 117.38 ± 0.05°. The only thermodynamic data for $SrThO_3$(cr) are measurements of the pressure of Sr(g) obtained by heating $SrThO_3$(cr) in a tungsten cell from 1677 to 2419 K by weight-loss Knudsen effusion (Ali (Basu) et al., 2001). The relevant reactions were thus assumed to be:

$5SrThO_3(cr) + W(cr) \leftrightarrow Sr_2WO_5(cr) + 5ThO_2(cr) + 3Sr(g)$ 1677 to 2047 K

and

$4SrThO_3(cr) + W(cr) \leftrightarrow SrWO_4(l) + 4ThO_2(cr) + 3Sr(g)$ 2138 to 2419 K.

Measurements were also made below 1670 K, but the establishment of the equilibrium (probably with $Sr_3WO_6(cr)$) was too slow for the results to be reliable. Based on these results calculated Gibbs energy of reaction from oxides: $SrO(cr) + ThO_2(cr) \rightarrow SrThO_3(cr)$

$\Delta_r G = - 3955 - 1.300T\ J{\cdot}mol^{-1}$.

The structure of $BaThO_3(cr)$ is a distortion of the perovskite structure with $a = 6.345 \pm 0.002$ Å, $b = 6.376 \pm 0.002$ Å and $c = 8.992 \pm 0.002$ Å. There is measurement of the Gibbs energy of formation of $BaThO_3(cr)$. (Mishra *et al.*, 1999) have studied the decomposition pressure of BaO(g) from the reaction $BaThO_3(cr) \rightarrow ThO_2(cr) + BaO(g)$ by Knudsen effusion from 1770 to 2136 K $\Delta_r G = - 21420 - 3.058T\ J{\cdot}mol^{-1}$.

Ternary Oxides

Thorium Carbonate Compounds

The following thorium carbonate solid phases have been synthesised in corresponding bicarbonate or carbonate solutions and characterised by chemical analysis, thermogravimetry and X-ray diffraction:

- $Na_6[Th(CO_3)_5]{\cdot}12H_2O(cr)$,
- $K_6[Th(CO_3)_5]{\cdot}xH_2O(cr)$, with $x = 8$ or 12,
- $(NH_4)_6[Th(CO_3)_5]{\cdot}3H_2O(cr)$,
- $[C(NH_2)_3]_6[Th(CO_3)_5]{\cdot}4H_2O(cr)$.

Thermodynamic data are not reported for these solids. However, the solubility data determined with $Na_6[Th(CO_3)_5]{\cdot}12H_2O(s)$ allow the determination of the solubility constant at zero ionic strength and the molar standard Gibbs energy of formation:

$\Delta_f G°(Na_6[Th(CO_3)_5]{\cdot}12H_2O, cr, 298.15\ K) = -8002.6 \pm 7.3\ kJ{\cdot}mol^{-1}$.

Thorium Phosphate Compounds

The temperature dependence of heat capacity of $KTh_2(PO_4)_3$ has been measured in the range from 5.5 to 350 K (A.V. Knyazev et al., 2014) by precision adiabatic vacuum calorimetry (figure 2). The experimental data were used to calculate standard thermodynamic functions, namely the heat capacity $C_p°(T)$, enthalpy $H°(T) - H°(0)$, entropy $S°(T) - S°(0)$ and Gibbs function $G°(T) - H°(0)$, for the range from $T \to 0$ to 350 K. The value of the fractal dimension D in the function of multifractal generalization of Debye's theory of the heat capacity of solids was estimated and the character of heterodynamics of structure was detected. The heat capacity of this substance in interval from 10 to 350 K gradually increases with rising temperature and does not show any peculiarities, but over the range from 5.5 to 10 K there is seen an anomalous heat capacity.

High-Temperature X-Ray Diffraction Investigation Method

The most objective information about thermal behavior of a compound in a wide range of temperatures may be acquired using *in situ* methods of investigation. First such method is thermoroentgenography or high-temperature XRD. But to get more adequate result it should be used with differential scanning calorimetry. These methods allow investigating different types of processes: polymorphic transitions, synthesis and decomposition of solid compounds, crystallization from the amorphous phase, *etc.*

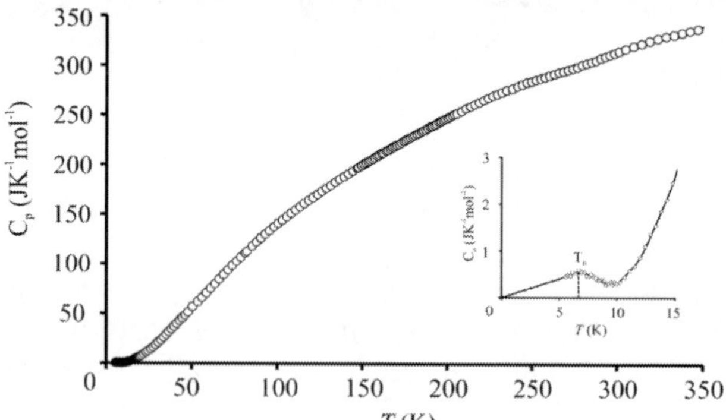

Figure 2. Temperature dependence of heat capacity of $KTh_2(PO_4)_3$.

Directly the experiment is as follows. In high-temperature XRD different kind of difractometer can be used (Fischer, Bardhan, & Geiger, 1983; Miyazaki et al., 2009; Panneerselvam, Krishnan, Nagarajan, & Antony, 2011; Tonegawa et al., 2010). The main condition is the presence of specific heating and cooling attachments. For example powder difractometer LabX XRD-6100 (Shimadzu) can be equipped by the attachment HA-1001 (Shmadzu) for heating of the samples in the temperature range 298-1173 K or attachment TTK-450 (Anton Paar) for cooling to 153 K (Figure 3). The result of the experiment is a series of XRD patterns of the substance under study which are obtained at selected temperatures. Depending on the type of structure is sometimes necessary to vary the conditions of recording X-ray and an algorithm for its processing (Figure 4). Than using procedure of analytical indexing of powder XRD patterns thermal dependencies of unit-cell parameters can be calculated (Kovba & Trunov, 1976; Porai-Koshitz, 1989). Such dependencies (also temperature dependencies of some structural parameters (atomic coordinates, bond lengths, bond angles, *etc.*) may be used) can be approximated by the polynomial, which in general can be written as follows: $l(T) = p_0 + p_1T + p_2T^2 + p_3T^3 + \ldots$ and can be used as a method of describing of thermal deformations. But this form of describing thermal deformations is rather cumbersome. That is why thermal expansion coefficients are commonly used. For example for unit-cell parameter *l*

$$\alpha_l = \varepsilon_l/dT = (dl/l)/dT,$$

where l – volume of the parameter; α_l – thermal expansion of l; ε_l – deformation of l parameter; T – temperature. So, thermal expansion coefficient characterizes temperature dependence of deformation of the parameter (S.K. Filatov, 1990).

There are a few ways to calculate thermal expansion coefficients (Chen et al., 2011; Hovis et al., 2008; *International Tables for Crystallography*, 2003). But the most convenient is using DTC (Deformation Tensor Computing) program, which was developed at St. Petersburg State University by R.I. Belousov and S.K. Filatov (Belousov & Filatov, 2007; R. S. Bubnova, Firsova, & Filatov, 2013). The initial data for thermal expansion calculating are the coefficients of the polynomials approximating the temperature dependence of the linear (and angular if necessary) unit cell parameters, temperature of interest, and unit cell symmetry (Figure 5). After the calculation the user get a table with linear and volume thermal expansion coefficients at the desired temperatures.

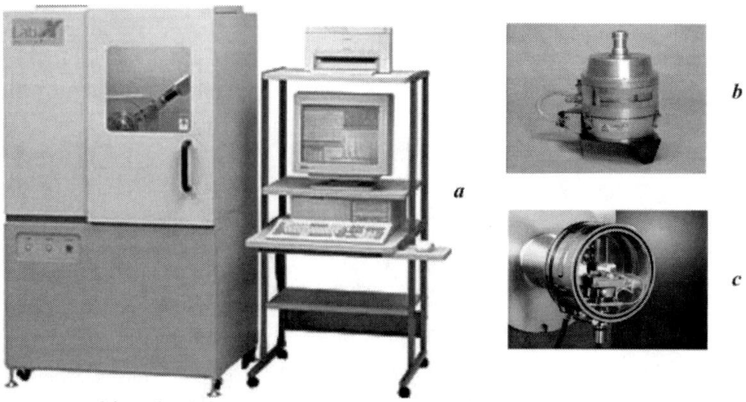

(Sources: www.shimadzu.com; www.anton-paar.com).

Figure 3. Powder difractometer LabX XRD-6000 (Shumadzu) (*a*), heating attachment HA-1001 (Shimadzu) (*b*) and cooling attachment TTK 450 (Anton Paar) (*c*).

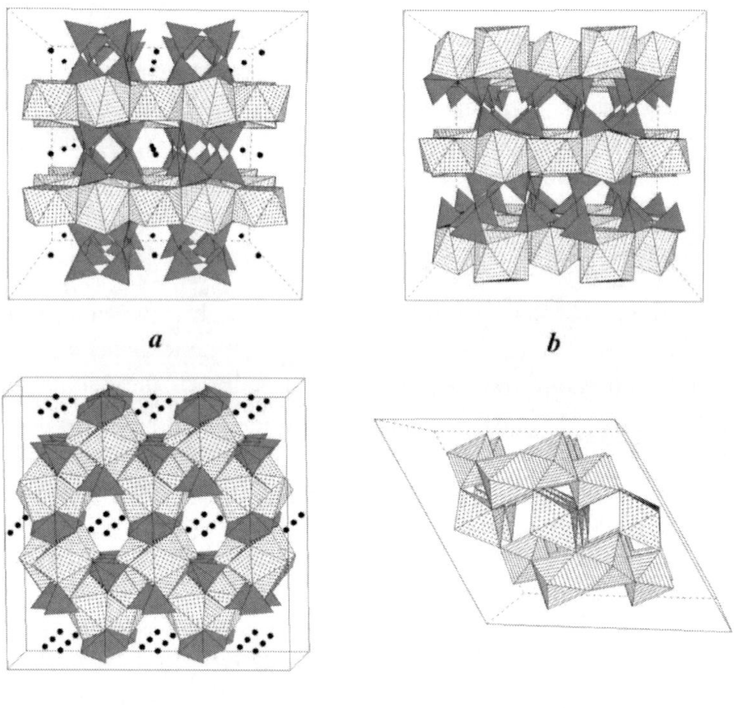

Figure 4. Crystal structures of some thorium compounds (*a* - $KNaCaTh(Si_8O_{20})$ (turkestanite); *b* - $Ca_2Th(Si_8O_{20})$ (ekanite); *c* - $KTh_2(PO_4)_3$; *d* - $ThTi_2O_6$ (brannerite).

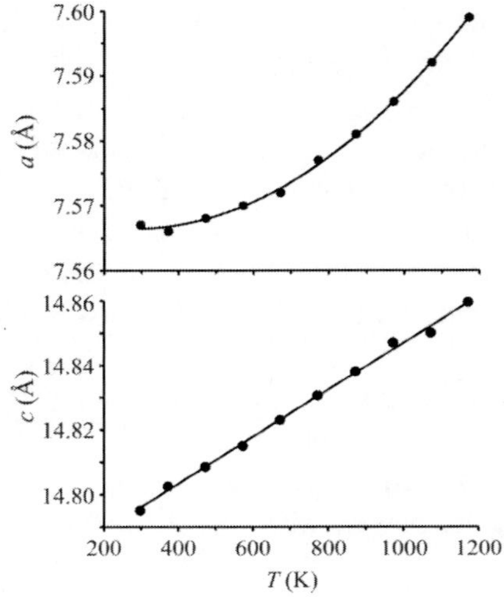

Figure 5. Temperature dependence of unit cells of $RbNaCaTh(Si_8O_{20})$.

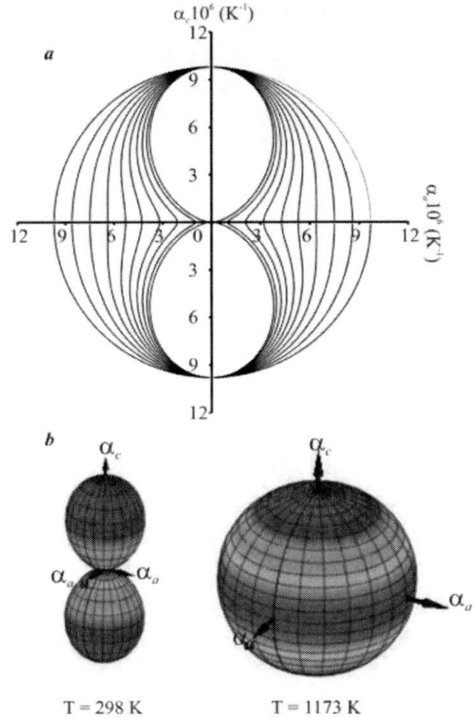

Figure 6. Thermal expansion diagrams (a - 2D; b – 3D) for $RbNaCaTh(Si_8O_{20})$.

For obtaining more evident information about high-temperature characteristics of the compound's deformation thermal expansion diagram are using. These figures constructing at different temperatures demonstrate priority directions of thermal expansion and its anisotropy. Value of thermal expansion coefficient in given direction corresponds to length of radius-vector which is traced from origin of coordinates to edge of figure of expansion (Figure 6). For the construction 3D version of the thermal expansion diagram DTP (Deformation Tensor Plot) program can be used (Belousov & Filatov, 2007; R. S. Bubnova, et al., 2013). But in 2013 in Lobachevsky State University was made another algorithm for Maplesoft© (Figure 6). Firstly the diagrams constructed in it were published in (A. V. Knyazev, Komshina, Zhidkov, & Plesovskikh, 2013). 2D-section of such diagrams made in different crystallographic planes may be also used to describe thermal expansion.

THERMAL EXPANSION OF DIFFERENT COMPOUNDS OF THORIUM

The most studied compound in thermal expansion is thorium oxide (Table 3). In the 2010th the amount of the thermal expansion data markedly increased due to the work performed on the basis of Lobachevsky State University (Nizhni Novgorod, Russia) and here we present a systematic analysis of data reported in the literature. Thermal deformations as any other physical characteristics of the compound are determined by the structure.

In $KNaCaTh(Si_8O_{20})$ (turkestanite) and The expansion along crystallographic direction *c* for both structures is considerably larger than along direction *a*, but the anisotropy decreases with an increase in temperature. The α_c/α_a ratio at T= 298 K is 4.8 and 8.6 for turkestanite and ekanite, respectively, whereas at T= 1173 K for both structures it is equal to 1.8. The anisotropic character of thermal deformations is due to structural features. As expected, the expansion in the layer, i.e., along direction *a*, is weaker because of the larger number of chemical bonds involved in the process, whereas the expansion in direction cinvolves only a small number of Si–O bonds linking the layers with each other. Despite the fact that both compounds have similar formula and crystallize in the same system, their thermal expansion coefficients are essentially different.

Table 3. The thermal expansion coefficients of thorium oxigen-containing compounds

Compound	Syngony	T, ΔT	$\alpha_a \cdot 10^6$, K^{-1}	$\alpha_b \cdot 10^6$, K^{-1}	$\alpha_c \cdot 10^6$, K^{-1}	$\alpha_\beta \cdot 10^6$, K^{-1}	Reference
ThO_2	cubic	300-673	7.1				(Samsonov, 1978)
		300-873	7.7				
α-$ThSiO_4$	tetragonal	298	2.2		8.5		new data
		373	2.4		9.8		
		473	2.7		11.6		
		573	3.0		13.3		
		673	3.3		15.0		
		773	3.6		16.7		
		873	3.8		18.4		
		973	4.1		20.2		
		1073	4.4		21.9		
		1173	4.7		23.6		
β-$ThSiO_4$	monoclinic	298	3.9	2.9	3.4	-1.4	new data
		373	4.4	3.5	3.4	-1.2	
		473	5.1	4.1	3.5	-1.0	
		573	5.8	4.8	3.6	-0.7	
		673	6.5	5.5	3.7	-0.4	
		773	7.2	6.2	3.8	-0.2	
		873	7.8	6.9	3.9	0.1	
		973	8.5	7.6	4.0	0.4	
		1073	9.2	8.3	4.1	0.6	
		1173	9.9	9.0	4.1	0.9	
$Ca_2Th(Si_8O_{20})$	tetragonal	298	0.2		1.4		(A.V. Knyazev, et al., 2012)
		373	0.2		1.2		
		473	0.3		1.0		
		573	0.5		1.1		
		673	0.8		1.4		
		773	1.1		1.9		
		873	1.5		2.6		
		973	2.0		3.5		
		1073	2.6		4.5		
		1173	3.3		5.8		
$LiNaCaTh(Si_8O_{20})$	cubic	298-1173	5.2				new data
$Na_2CaTh(Si_8O_{20})$	cubic	298	3.0				new data
		373	3.5				
		473	4.1				
		573	4.7				
		673	5.2				
		773	5.8				
		873	6.4				
		973	7.0				
		1073	7.6				
		1173	8.2				

Table 3. (Continued)

Compound	Syngony	T, ΔT	$\alpha_a \cdot 10^6$, K^{-1}	$\alpha_b \cdot 10^6$, K^{-1}	$\alpha_c \cdot 10^6$, K^{-1}	$\alpha_\beta \cdot 10^6$, K^{-1}	Reference
$KNaCaTh(Si_8O_{20})$	tetragonal	298	1.7		8.3		(A.V. Knyazev, et al., 2012)
		373	2.1		8.5		
		473	2.5		8.8		
		573	3.0		9.0		
		673	3.5		9.3		
		773	3.9		9.5		
		873	4.4		9.8		
		973	4.8		10.0		
		1073	5.3		10.3		
		1173	5.7		10.6		
$RbNaCaTh(Si_8O_{20})$	tetragonal	298	0.2		9.8		new data
		373	1.0		9.8		
		473	2.1		9.8		
		573	3.2		9.8		
		673	4.3		9.8		
		773	5.3		9.8		
		873	6.4		9.8		
		973	7.5		9.8		
		1073	8.6		9.8		
		1173	9.6		9.8		
$CsNaCaTh(Si_8O_{20})$	tetragonal	298	0,7		36,9		new data
		373	0,9		33,9		
		473	1,2		29,8		
		573	1,4		25,8		
		673	1,7		21,2		
		773	1,9		17,9		
		873	2,1		14,0		
		973	2,4		10,1		
		1073	2,6		6,2		
		1173	2,9		2,4		
$Ca_6Th_4(SiO_4)_6O_2$	tetragonal	298	7.6		5.2		new data
		373	8.0		5.7		
		473	8.5		6.4		
		573	9.1		7.1		
		673	9.6		7.9		
		773	10.1		8.6		
		873	10.7		9.3		
		973	11.2		10.0		
		1073	11.7		10.7		
		1173	12.2		11.4		
$Th_4(PO_4)_4(P_2O_7)$	orthorhombic	298	0.9	0.1	1.8		(A.V. Knyazev, et al., 2011)
		373	1.1	0.3	2.2		
		473	1.4	0.6	2.8		
		573	1.7	1.0	3.4		

Compound	Syngony	T, ΔT	$\alpha_a \cdot 10^6$, K^{-1}	$\alpha_b \cdot 10^6$, K^{-1}	$\alpha_c \cdot 10^6$, K^{-1}	$\alpha_\beta \cdot 10^6$, K^{-1}	Reference
		673	2.1	1.3	4.0		
		773	2.4	1.6	4.6		
		873	2.7	1.9	5.1		
		973	3.0	2.3	5.7		
		1073	3.3	2.6	6.3		
		1173	3.7	2.9	6.9		
$KTh_2(PO_4)_3$	monoclinic	298-1173	7.7	6.7	13.3	-0.9	(A.V. Knyazev, et al., 2011)
$CaTh(PO_4)_2$	monoclinic	298-1173	12.0	9.9	10.9	0.7	(A.V. Knyazev, et al., 2011)
β-$PbTh(VO_4)_2$	tetragonal	298	3.03		2.4		(A.V. Knyazev, et al., 2011)
		373	2.3		3.2		
		473	1.8		4.2		
		573	2.0		5.2		
		673	2.7		6.2		
		773	4.1		7.3		
		873	6.1		8.3		
		973	8.8		9.3		
		1073	12.0		10.3		
		1173	15.9		11.3		
$BaTh(VO_4)_2$	monoclinic	298	6.9	4.4	10.0	1.1	(A.V. Knyazev, et al., 2011)
		373	7.8	5.2	11.2	1.1	
		473	9.1	6.4	12.7	1.1	
		573	10.4	7.5	14.2	1.1	
		673	11.7	8.6	15.7	1.1	
		773	12.9	9.8	17.2	1.1	
		873	14.2	10.9	18.7	1.1	
		973	15.5	12.0	20.2	1.1	
		1073	16.7	13.1	21.6	1.1	
		1173	18.0	14.3	23.1	1.1	

In the more closely packed structure of $KNaCaTh(Si_8O_{20})$ (Figure 4a), they are higher, because an increase in the amplitude of atom vibrations inevitably leads to an increase in the unit cell parameters, whereas the presence of "voids" in the structure of $Ca_2Th(Si_8O_{20})$ can level out the effect of temperature because of the shift of vibrations toward these voids.

Thorium phosphate diphosphate ($Th_4(PO_4)_4(P_2O_7)$) at 298 K exhibits large anisotropy of the thermal expansion coefficients (α_a: α_b: α_c= 22 : 1 : 45), but with an increase in temperature the difference between the thermal expansion coefficients decreases (α_a: α_b: α_c= 1.2 : 1 : 2.4 at T= 1173 K), which leads to thermal stability of the structure. The largest thermal deformations are observed along crystallographic direction ccoinciding with the P–O–P bridging bonds of the pyrophosphate group which, according to the structural data, are the longest among all the P–O bonds in this structure.

The compounds $KTh_2(PO_4)_3$ and $CaTh(PO_4)_2$, as seen from Table 3, exhibit insignificant anisotropy of thermal expansion and are highly expanding. The monoclinic angle in the structure of these compounds does not noticeably vary with temperature. Among the examined compounds, the smallest anisotropy of thermal expansion is observed in brabantite $CaTh(PO_4)_2$.

It is clear from the data listed in Table 3 that an insignificant anisotropy of thermal expansion is observed in β-$PbTh(VO_4)_2$ crystals with zircon structure. This phase can be ranked among the phases of moderate expansion. On the contrary, a significant anisotropy of thermal expansion is observed in $BaTh(VO_4)_2$ crystals with monazite structure, which is attested by the ratios $\alpha_c:\alpha_a = 1.47$ and $\alpha_c:\alpha_b = 2.29$ (much greater than unity). However, it should be noted that there is a tendency toward leveling out the thermal expansion coefficients with increasing temperature; these ratios become $\alpha_c:\alpha_a = 1.28$ and $\alpha_c: \alpha_b = 1.62$ at T= 1173 K.

CONCLUSION

In this chapter, the analysis is made of the most important thermal and thermodynamic properties of thorium oxide compounds. Unfortunately, in the literature there is no reliable and complete information on complex compounds of thorium. However, this information is essential for materials scientists working with thorium materials.

The reported study was supported by RFBR, research project No. 13-03-00152 A.

REFERENCES

Ackermann, R. J., Rauh, E. G., Thorni, R. J. & Cannon, M. C. (1963). A thermodynamic study of the thorium: oxygen system at high temperatures. *The Journal of Physical Chemistry, 67*, 762-769.

Ali (Basu), M., Mishra, R., Bharadwaj, S. R., Kerkar, A. S., Dharwadkar, S. R. & Das, D. (2001). Thermodynamic stability of solid $SrThO_3$. *J. Nucl. Mater., 299*, 165-170.

Allen, K.A. & McDowell, W.J. (1963). The thorium sulfate complexes from di-n-decylamine sulfate extraction equilibria. *The Journal of Physical Chemistry, 1963, 67(5)*, 1138-1140.

Bagnall, K. W., Brown, D. & Easy, J. F. (1968). Preparation, crystallographic properties and infrared spectra of quadrivalent oxyhalides of thorium, protactinium, uranium, and neptunium. *Journal of the Chemical Society A, 2*, 288-291.

Bartel, F. H., Dahlkamp, F. J., Ditz, R., Sarbas, B., Schubert, P. & Topper, W. (1991). Minerals (silicates). Deposits. Mineral Index. *Gmelin handbook of inorganic and organometallic chemistry, Thorium, Supplement A1b*, 440.

Baskin, Y., Harada, Y. & Handwerk, J. H. (1961). Crystalline thorium borate. *Journal of the American Ceramic Society, 44(9)*, 456-459.

Belousov, R. I., & Filatov, S. K. (2007). Algorithm for calculating the thermal expansion tensor and constructing the thermal expansion diagram for crystals. *Glass Physics and Chemistry, 33*(3), 271-275.

Bilinski, H. & Ingri, N. (1967). *Croatica chemica acta, 39*, 235.

Bressat, R., Claudel, B. & Trambouze, Y. (1963). Relations disomorphisme dans la série des oxalates hydrates de thorium et d'uranium. *Journal de chimie physique et de physico-chimie biologique, 60(11-2)*, 1265.

Bubnova, R. S., Firsova, V. A., & Filatov, S. K. (2013). Software for determining the thermal expansion tensor and the graphic representation of its characteristic surface (theta to tensor-TTT). *Glass Physics and Chemistry, 39*(3), 347-350.

Burkhart, M. J. & Thompson, R. C. (1972). Assotiation of chromium(VI) with neptunium(IV) and thorium(IV) in perchlorate solution. Journal *of the American Chemical Society, 94(9)*, 2999-3002.

Busch, G., Marincek, O., Menth, A. & Vogt, O. (1965). Magnetic properties of ordering rare-earth antimonides. *Physics letters (Netherlands), 14(4)*, 262-264.

Bushuev, N. N. & Trunov, V. K. (1975). *Russian Journal of Inorganic Chemistry*, 20, 1143.

Chauvenet, E. (1911). Contribution à l'étude des dérivés halogénés et oxy-halogénés du thorium. *Annales de chimie et de physique, 23(8)*, 425-490.

Cheda, J. A. R., Westrum, E. F., Morss, L. R. (1976). Heat capacity of $Th(NO_3)_4 \cdot 5H_2O$ from 5 to 350 K. *Journal of Chemical Thermodynamics, 8*, 25-29.

Chen, X., Zhou, H. D., Kiswandhi, A., Miotkowski, I., Chen, Y. P., Sharma, P. A., et al. (2011). Thermal expansion coefficients of Bi_2Se_3 and Sb_2Te_3 crystals from 10 K to 270 K. *Applied Physics Letters, 99*(26).

Cox, J. D., Wagman, D. D. & Medvedev, V. A. (1989). CODATA Key Values for Thermodynamics. *Hemisphere Publ. Corp., New York*, 27.

Danesi, P. R., Magini, M., Margherita, S. & D'Alessandro, G. (1968). Energia nucleare, 15(5), 335.

Danon, J. (1960). Determination of the stability constants of thorium nitrate complexes with anion-exchange resins. *Journal of Inorganic and Nuclear Chemistry, 13(1-2)*, 112-118.

Darnell, A. J., McCollum, W. A. & Milne, T. A. (1960). Vapor pressure of thorium. *The Journal of Physical Chemistry, 64(3)*, 341-346.

Day, R. A. & Stoughton, R. W. (1950). Chemistry of thorium in aqueous solutions. I. Some organic and iInorganic complexes. *Journal of the American Chemical Society, 72(12)*, 5662-5666.

D'Eye, R. W. M., Sellman, P. G. & Murray, J. R. (1952). The thorium–selenium system. *Journal of the Chemical Society (Resumed)*, 2555-2562.

Ditz, R., Sarbas, B. & Schubert, P. (1990). Natural Occurrence Minerals (excluding silicates). *Gmelin handbook of inorganic and organometallic chemistry. Thorium, Supplement A1a*, 391.

Dodgen, H. W. & Rolefson G. K. (1949). The complex ions formed by iron and thorium with fluoride in acid solution. *Journal of the American Chemical Society, 71*, 2600-2607. Finkelnburg, W.

Ferraro, J. R., Katzin, L. L. & Gibson, G. (1956). *Journal of Inorganic and Nuclear Chemistry, 2*, 118.

Filatov, S. K. (1990). *High-temperature crystal chemistry. Theory, methods and research results*. Leningrad: "Subsoils".

Fischer, G. R., Bardhan, P., & Geiger, J. E. (1983). The lattice thermal-expansion of hydroxyapatite. *Journal of Materials Science Letters, 2*(10), 577-578.

Fontenau, G. & L'Helgoulach, Lucas J. (1977). *Materials Research Bulletin, 12*, 25.

Freundlich, W., Erb, A. & Pages M. (1974). Double Th and Np(IV) orthovanadates and arsenates with alkali metals V-As substitution on tetrahedral sites. *Rev. Chim. Miner., 11*, 598-606.

Gayer, K. H. & Leider, H. (1954). The solubility of thorium hydroxide in solutions of sodium hydroxide and perchloric acid at 25°. *Journal of the American Chemical Society, 76(23)*, 5938-5940.

Gingerich, K. A. (1969). *High Temperature Science, 1*, 258.

Glushko, V. P. (1965-1982). Thermal Constants of Substances. Moscow: VINITI, 1-10.

Gordon, J. E., Montgomery, H., Noer, R. J., Pickett, G. R. & Tobón, T. (1966). Superconductivity of thorium and uranium. *Phys. Rev., 152*, 432-437.

Griffel, M. & Skochdopole, R. E. (1953). The heat capacity and entropy of thorium from 18 to 300 K. *J. Am. Chem. Soc., 75*, 5250-5251.

Gschneidner, K. A., Elliott, R. O. & McDonald, R. R. (1962). *Journal of the Physics and Chemistry of Solids, 23,* 1201.

Harari, A., Thery, J. & Collongues, R. (1967). *Revue internationale des hautes temperatures et des refractaires, 4*, 207.

Harris, L. A. & Finch, C. B. (1972). Phase stability in the ThGeO sub 4 system in the range 750-1450 degrees C at 1 ATM. *The American Mineralogist, 57(11-12)*, 1894-1898.

Hildenbrand, D. L. & Murad, E. (1974). Mass spectrometric studies of gaseous ThO and ThO_2. *The Journal of Chemical Physics, 61(3)*, 1232-1237.

Hoch, M. & Johnston, H. L., (1961). The heat capacity of aluminum oxide from 1000 to 2000° and of thorium dioxide from 1000 to 2500°. *The Journal of Physical Chemistry, 65*, 1184-1185.

Huber Jr. E. J., Holley Jr. & C. E., Meierkord, E.H. (1952). The heats of combustion of thorium and uranium. *J. Am. Chem. Soc., 74*, 3406-3408.

Humbach, W. (1955). Ionisierungsenergien von Atomen und Atomionen. *Naturwissenschaften, 42*, 35-37.

International Tables for Crystallography. (2003). (Vol. D. Physical properties of crystal). Dordrecht/Boston/London: Kluwer Academic Publishers.

Keller, C. (1965). Die reaktion der dioxide der elemente thorium bis americium mit niob- und tantalpentoxid. *Journal of Inorganic and Nuclear Chemistry (Англия), 27(6)*, 1233-1246.

Klinkenberg, P. F. A. (1950). Analysis of the spectrum of doubly ionized thorium. *Physica, 16(7-8)*, 618-644, 645-650.

Knyazev, A. V., Chernorukov, N. G. & Komshina, M. E. (2012). Synthesis and study of some thorium-containing silicates of the ekanite group. *Russian Radiochemistry, 54(5)*, 431-434.

Knyazev, A. V., Chernorukov, N. G., Sazonov, A. A. & Bulanov, E. N. (2011). Synthesis and study of compounds of the general formula $M^{II}Th(VO_4)_2$ (MII – Mn, Cd, Ca, Sr, Pb, Ba). *Russian Journal of Inorganic Chemistry, 57 (5)*, 639-645.

Knyazev, A. V., Chernorukov, N. G., Sazonov, A. A. & Komshina, M. E. (2011). High-temperature and spectroscopic study of thorium phosphates. *Russian Radiochemistry, 53(6)*, 587–592.

Knyazev, A. V., Komshina, M. E., Zhidkov, A. V., & Plesovskikh, A. S. (2013). Crystal structure and thermal expansion of $RbNaCaTh(Si_8O_{20})$. *Russian Journal of Inorganic Chemistry, 58*(10), 1172-1176.

Knyazev, A. V., Smirnova, N. N., Manyakina, M. E. & Shushunov A.N. (2014). Low-temperature heat capacity and thermodynamic functions of $KTh_2(PO_4)_3$. *Thermochimica Acta*, doi:10.1016/j.tca.2014.03.033

Koppel, J. (1910). *Zeitschrift fuer anorganische und allgemeine Chemie, 67*, 293.

Kovba, L. M., & Trunov, V. K. (1976). *X-ray analysis*. Moscow: Moscow State University.

Kraus, K. A. & Holmberg, R. W. (1954). Hydrolytic behavior of metal Ions. III. Hydrolysis of thorium(IV). *The Journal of Physical Chemistry, 58(4)*, 325-330.

Krylov, E. I. & Chukhlantsev, V. G.. (1957). *Journal of Analytical Chemistry, 12*, 451.

Kusakabe, T. & Imoto, S. (1972). *Technology Reports Osaka University, 22*, 477.

Marshall, W. L., Gill, J. S. & Secoy, C. H. (1951). The thorium nitrate-water system above 20°. *Journal of the American Chemical Society, 73(10)*, 4991-4992.

Matijevic, E., Abramson, M. B., Schulz, K. F. & Kerker, M. (1960). Detection of metal ion hydrolysis by coagulation. *The Journal of Physical Chemistry, 64*, 1157-1161.

Mayer, S. W.,Owens, B. B., Rutherford, T. H. & Serrins, R. B. (1960). High-temperature free energy, entropy, enthalpy and heat capacity of thorium sulfate. *The Journal of Physical Chemistry, 64(7)*, 911-914.

Mayorova E. P. & Fomin V. V. (1958). *Russian Journal of Inorganic Chemistry*, 3, 1937.

Mayorova E. P. & Fomin V. V. (1959). *Russian Journal of Inorganic Chemistry*, 4, 2511.

Mazeina, L., Ushakov, S. V., Navrotsky, A. & Boatner, L. A. (2005). Formation enthalpy of $ThSiO_4$ and enthalpy of the thoriteohuttonite phase transition. *Geochim. Cosmochim. Acta, 69*, 4675-4683.

Mishra, R., Ali (Basu), M., Bharadwaj, S. R., Kerkar, A. S., Das, D. & Dharwadkar, S. R. (1999). Thermodynamic stability of barium thorate, $BaThO_3$, from a Knudsen effusion study. *J. Alloys Compd. 290*, 97-102.

Miyazaki, H., Ushiroda, I., Itomura, D., Hirashita, T., Adachi, N., & Ota, T. (2009). Thermal expansion of hydroxyapatite between-100 degrees C and

50 degrees C. *Materials Science & Engineering C-Biomimetic and Supramolecular Systems, 29*(4), 1463-1466.

Murthy, P. R. & Patel, C. C. (1963). Crystal structure of thorium perchlorate tetrahydrate by X-ray method. *Journal of Inorganic and Nuclear Chemistry, 25(3)*, 310-312.

Nakamura, J., Takahashi, Y., Izumi, S. & Kanno, M. (1980). Heat capacity of metallic uranium and thorium from 80 to 1000 K. *J. Nucl. Mater., 88*, 64-72.

Osborne, D. W. & Westrum Jr. E. F. (1953). The heat capacity of thorium dioxide from 10 to 305 K. The heat capacity anomalies in uranium dioxide and neptunium dioxide. *J. Chem. Phys., 21*, 1884-1887.

Panneerselvam, G., Krishnan, R. V., Nagarajan, K., & Antony, M. P. (2011). Heat capacity and thermal expansion of samarium titanate. *Materials Letters, 65*(12), 1778-1780.

Porai-Koshitz, M. A. (1989). *Basic of structural analisys of chemical compounds*. Moscow: High School.

Portanova, R., Bernardo, P. D., Traverso, O., Mazzocchin, G. A. & Magon, L. (1975). *Journal of Inorganic and Nuclear Chemistry, 37*, 2177.

Portanova, R., Tomat, G., Cassol, A. & Magon, L. (1972). Equilibria of thorium(IV) with monocarboxylate ligands. *Journal of Inorganic and Nuclear Chemistry, 34(5)*, 1685-1690.

Rand, M., Fuger, J., Grenthe I., Neck, V. & Rai, D. (2008). Chemical thermodynamics of thorium. *OECD NEA Publishing*, 942.

Revcolevschi, A. (1970) *Revue internationale des hautes temperatures et des refractaires (Франция), 7*, 73.

Samsonov, G. V. (1978). Physico-chemical properties of oxides: critical table. *Moscow: Metallurgiya*, 472.

Scaife, D. E., Turnbull, A. G. & Wylie, A. W. (1965). Thorium oxyiodide. *Journal of the Chemical Society(Resumed)*, 1432-1437.

Schuiling, R. D., Vergonwen, L. & van der Rijst, H. (1976). Gibbs energies of formation of zircon ($ZrSiO_4$), thorite($ThSiO_4$) and phenasite ($BeSiO_4$). *The American Mineralogist, 61*, 166-168.

Sergeev G. M., Astrashkova L. G. & Yagodinskaya N. N. (1970). *Russian Radiochemistry, 12*, 392.

Sood, R. R. & Stage, R. A. (1966). Pressure- induced dehydration reactions and transitions in inorganic hydrates. *Science, 154(3747)*, 388-390.

Stoughton, R. W., Fry, A. J. & Barney, J. E. (1961). A study of thorium-iodate complexing in aqueous solutions. *Journal of Inorganic and Nuclear Chemistry, 19(3-4)*, 286-297.

Sugar, J. (1973). Ionization energies of neutral actinides. *The Journal of Chemical Physics, 59(2)*, 788-791.

Tedesco, P. H. & Anon, M. C. (1972). Thorium-acetate and thorium-propionate complexes. *Journal of Inorganic and Nuclear Chemistry, 34(7)*, 2271-2276.

Tonegawa, T., Ikoma, T., Suetsugu, Y., Igawa, N., Matsushita, Y., Yoshioka, T., et al. (2010). Thermal expansion of type A carbonate apatite. *Materials Science and Engineering B-Advanced Functional Solid-State Materials, 173*(1-3), 171-175.

Thoret, J. (1971). 3 systems - Th(MoO4)2-Li2MoO4 - Th(MoO4)2-Na2MoO4 - Th(MoO4)2-K2MoO4. C. R. *Acad. Sci. Ser.*, C273, 1431-1434.

Wallace, D. C. (1960). Specific heat of thorium at high temperatures. *The Physical Review 120*, 84-88.

Westrum, E. F. & Gronvold, F. (1962). Chemical thermodynamics of the actinide element chalcogenides. *Thermodynamics of nuclear materials, Proc. Symp. Held 21-25 May in Vienna,* 3-37.

Zielen, A. J. (1959). Thermodynamics of the sulfate complexes of thorium. *Journal of the American Chemical Society, 81(19)*, 5022-5028.

In: Thorium
Editor: Renae Jamison
ISBN: 978-1-63321-309-8

Chapter 8

UNIQUE ROLE OF THORIUM IN CLOSED (Th-U-Pu)-FUEL CYCLE DUE TO THE INVOLVEMENT OF HYBRID "FUSION-FISSION" REACTORS WITH TH-BLANKET TO GENERATE ADVANCED (^{231}Pa+^{232}U+^{233}U)-FUEL: LONG-LIFE POWER REACTOR CORES AND NON-PROLIFERATION

A. N. Shmelev, G. G. Kulikov*, E. G. Kulikov, V. A. Apse and N. I. Geraskin
Department of Theoretical and Experimental Reactor Physics
National Research Nuclear University
"Moscow Engineering Physics Institute", Russia

ABSTRACT

Technology of controlled thermonuclear fusion (CTF) is traditionally regarded as a practically inexhaustible energy source. However, development, mastering, broad deployment of fast breeder reactors and

* E-mail address: ggkulikov@mephi.ru.

closure of nuclear fuel cycle (NFC) can also extend fuel base of nuclear power industry (NPI) up to practically unlimited scales. Under these conditions, it seems reasonable to introduce into a circle of the CTF-related studies the works directed towards solving some principal problems which can appear in a large-scale NPI in closed NFC.

The first challenge is a large scale of operations in NFC back-end that should be reduced by achieving substantially higher fuel burn-up in power nuclear reactors. As was shown in [1], the use of ^{231}Pa-^{232}Th-^{232}U-^{233}U fuel in light-water reactor (LWR) opens a possibility of principle to reach very high (about 30% HM) or even ultra-high fuel burn-up.

The second challenge is a potential unauthorized proliferation of fissionable materials. As is known, a certain remarkable quantity of ^{232}U being introduced into uranium fraction of nuclear fuel can produce a serious barrier against switching the fuel over to non-energy purposes.

As is shown in publications [2, 3], involvement of hybrid thermonuclear reactors (HTR) into NPI structure can substantially facilitate resolving these problems. If HTR will be involved into NPI structure, then main HTR mission consists not in energy generation but in production of nuclear fuel with a certain isotope composition.

The present chapter analyzes some neutron-physical features in production of advanced nuclear fuels in thorium HTR blankets. The obtained results demonstrated that such a nuclear fuel may be characterized by very stable neutron-multiplying properties during full LWR operation cycle and by enhanced proliferation resistance too. The chapter evaluates potential benefits from involvement of HTR with thorium blanket into the international closed NFC.

1. Neutron-Physical Peculiarities of Nuclear Fuel Production in Thorium HTR Blanket

1.1. Comparison of Thorium and Uranium HTR Blankets

As is known, the HTR concept presumes the use of high-energy (14.1 MeV) neutrons emitted in (D,T) fusion reaction for irradiation of uranium or thorium blankets. Just neutron irradiation of thorium blanket allows the high-energy fusion neutrons to show more strikingly their unique capabilities in initiation of threshold (n,2n)- and (n,3n)-reactions. Micro cross-sections of these threshold neutron reactions with ^{232}Th are substantially larger than those with ^{238}U while fission cross-sections of ^{232}Th are remarkably smaller than those of ^{238}U (Figures 1, 2).

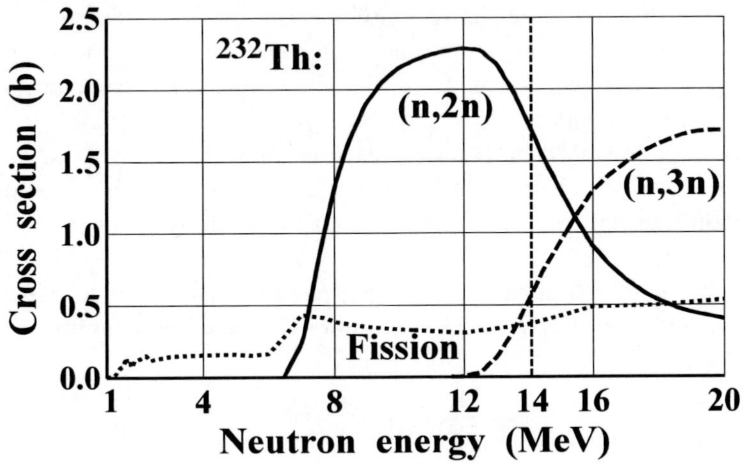

Figure 1. Energy dependence of fission cross-sections and threshold (n,2n)- and (n,3n)-reactions of ^{232}Th (JENDL-4.0).

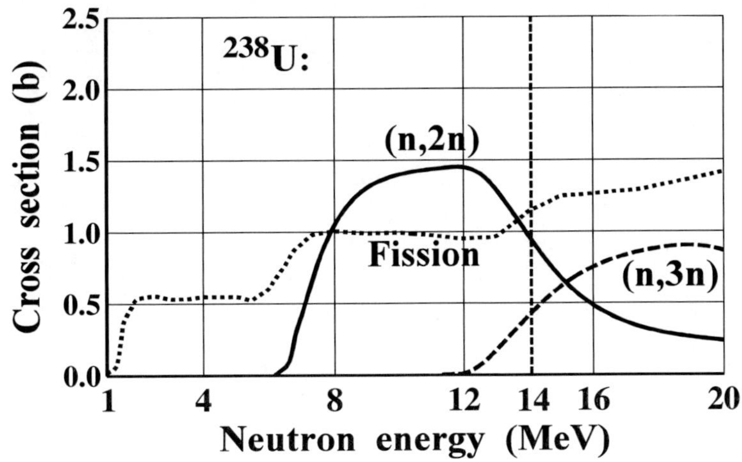

Figure 2. Energy dependence of fission cross-sections and threshold (n,2n)- and (n,3n)-reactions of ^{238}U (JENDL-4.0).

It may be concluded from energy dependencies of micro cross-sections for the threshold (n,2n), (n,3n) and (n,f) reactions of ^{232}Th and ^{238}U that both nuclides are able to breed fusion neutrons with a high efficiency. However, some difference consists in the fact that ^{232}Th multiplies fusion neutrons primarily through the threshold (n,2n) and (n,3n) channels while ^{238}U – mainly through fission channel. This means the lower energy generation in thorium

HTR blanket and the larger amounts of the isotopes produced through the threshold channels. If these isotopes are helpful for NFC, then they should be taken into account when comparing the future prospects on thorium and uranium energy utilization.

Some experimental data [4-7] on reaction rates induced by irradiation of ^{232}Th and ^{238}U by fusion neutrons are presented in Table 1. These data confirm interrelationships between micro cross-sections of neutron reactions.

Table 1. Reaction rates (per one 14.1-MeV fusion neutron) in experimental assemblies containing thorium and uranium

Reaction	Thorium experiments	Reaction*	Uranium experiments
^{232}Th(n,2n)^{231}Pa	0.42±0.04	^{238}U(n,2n)^{237}Np	0.277±0.008
^{232}Th(n,3n)^{230}Th	0.30±0.05	^{238}U(n,3n)^{236}U	0.327±0.052
^{232}Th(n,f)	0.174±0.01	^{238}U(n,f)	1.18±0.06
		^{235}U(n,f)*	0.281±0.017
^{232}Th(n,γ)^{233}U	1.63±0.10	^{238}U(n,γ)^{239}Pu	4.08±0.24
Neutron leakage	0.78±0.04	Neutron leakage	0.41±0.02

* Metallic natural uranium (0.71% ^{235}U) was used as a material to be irradiated in experimental assemblies [6, 7].

It follows from these experimental results (Table 1) that yield of (^{231}Pa+^{230}Th)-nuclides in thorium HTR blanket (in essentially fresh HTR blanket) will be larger than yield of (^{237}Np+^{236}U)-nuclides in uranium HTR blanket per one fission reaction (or per two nuclei of fission products). In addition, energy generation (per one (DT)-reaction in plasma) in thorium HTR blanket is lower by a factor of 8.6 than that in natural uranium HTR blanket.

Therefore, thorium blanket with relatively small energy generation rate may be characterized by relatively simpler design and better economic indicators. It can be important at resolving the problem whether it is economically better to generate energy in a hybrid HTR or in a nuclear power reactor. One else important circumstance consists in the fact that secondary fuel with large content of ^{233}U produced in thorium HTR blanket is the higher-quality fuel for thermal nuclear reactors as compared with plutonium-based fuel produced in uranium HTR blanket [8].

1.2. Chains of Nuclide Transformations in Thorium HTR Blanket

Under neutron irradiation of thorium the following chains of nuclide transformations can be initiated in HTR blanket (Fig. 3):

1. So called "traditional" branch that starts from radiative neutron capture by thorium (the capture channel):

- $^{232}Th(n,\gamma)^{233}Pa(\beta^{-},T_{1/2}=27\ d)^{233}U(n,\gamma)^{234}U(n,\gamma)$...

2. "Non-traditional" branch that starts from threshold neutron (n,2n) and (n,3n)-reactions with thorium (the threshold channels):

- $^{232}Th(n,2n)^{231}Th(\beta^{-},T_{1/2}=26\ h)^{231}Pa(n,\gamma)^{232}Pa(\beta^{-},T_{1/2}=1.3\ d)$ $^{232}U(n,\gamma)^{233}\ U(n,\gamma)^{234}U(n,\gamma)$...
- $^{232}Th(n,3n)^{230}Th(n,\gamma)^{231}Th(\beta^{-},T_{1/2}=26\ h)^{231}Pa(n,\gamma)^{232}U(n,\gamma)$ $^{233}U(n,\gamma)^{234}U(n,\gamma)$...

These reactions promote accumulation of isotopes ^{231}Pa and ^{232}U in the irradiated thorium, in addition to ^{233}U. As it follows from Table 1, rates of the threshold (n,2n)- and (n,3n)-reactions exceed 40% of radiative neutron capture reaction rate in thorium that leads to build-up of main fissile isotope ^{233}U.

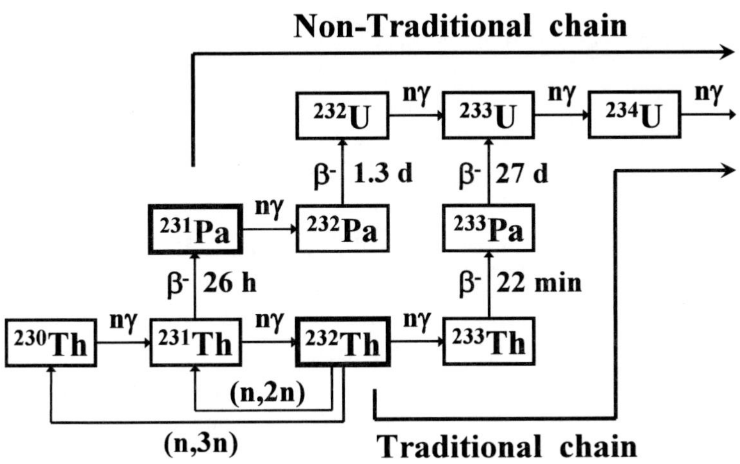

Figure 3. Chain of nuclide transformations in thorium blanket.

1.3. Nuclide Composition of Fuel Produced in Thorium HTR Blanket

It follows from Table 2 [9] that the longer irradiation time of thorium, the larger content of ^{230}Th and ^{231}Pa in thorium. If protactinium could be extracted from irradiated thorium by radiochemical technologies, then ^{230}Th remains in thorium, and multiple recycle of regenerated thorium in HTR blanket becomes reasonable for further accumulation of ^{231}Pa.

Table 2. Nuclide composition of fuel accumulated in thorium HTR blanket, kg/t

Nuclide	Irradiation time, days [1]				
	200	400	600	800	1000
^{230}Th	0.29	0.58	0.87	1.15	1.43
^{232}Th	997.	993.	989.	986.	982.
^{231}Pa	0.68	1.30	1.85	2.33	2.75
^{233}Pa [2]	0.45	0.46	0.46	0.47	0.47
^{232}U	0.03	0.13	0.29	0.51	0.78
^{233}U	1.77	3.94	6.09	8.23	10.3
^{234}U	0.0034	0.014	0.033	0.058	0.09
Fission products	0.51	4.54	6.87	9.27	11.64

[1] Neutron load on the first wall - 1 MW/m^2

[2] Nuclide ^{233}Pa through β-decay ($T_{1/2}$ = 27 days) transforms into ^{233}U

The ratios between accumulation rates of the threshold channel nuclides ^{230}Th, ^{231}Pa, ^{232}U and main fissile nuclide ^{233}U (with proper accounting for ^{233}Pa) together with relative contents of ^{232}U in uranium fraction of the produced fuel composition are presented in Table 3.

It is noteworthy here that products of (n,2n) and (n,3n) threshold reactions are accumulated in thorium blanket, firstly, without neutron consumption but even with some neutron breeding, and, secondly, total amount of these products is comparable with amount of main fissile nuclide.

Also, as it can be seen, ^{232}U content in produced uranium reaches significant values (Table 4 [9]). Such uranium fraction may be called as unique "highly enriched" uranium because it contains above 90% of fissile ^{233}U and unprecedented large fraction of ^{232}U (at the level of several percents!). Fractions of other, heavier uranium isotopes (^{234}U and ^{235}U) are extremely small. It is appropriate here to note that K.R. Schultz (Ref. 8) has already shown that isotope ^{232}U can enhance proliferation resistance of uranium fraction in the produced fuel composition. However, ^{232}U-induced radiation effects require applying distant equipment and remote technologies for nuclear fuel manufacturing. Previously, this fact was estimated as a very weighty argument in discussions on non-attractive production of highly radioactive fuel. Currently, when real NFC closure is near to realization, the closed NFC options with incomplete purification of spent nuclear fuel from highly radioactive products are being considered now from nuclear non-proliferation positions in a practical sense [10]. This factor appears in HTR blanket in a natural manner.

Table 3. Ratios between accumulation rates of the threshold channel nuclides ^{230}Th, ^{231}Pa, ^{232}U and main fissile nuclide ^{233}U (with accounting for ^{233}Pa) together with relative contents of ^{232}U in uranium fraction of the produced fuel composition

Nuclide	Irradiation time, days[1]				
	200	400	600	800	1000
$(^{230}Th + ^{231}Pa + ^{232}U) / (^{233}U + ^{233}Pa)$	0.451	0.457	0.458	0.459	0.461
$^{232}U / (^{232}U + ^{233}U + ^{233}Pa)$	0.015	0.029	0.042	0.055	0.067

[1] Neutron load on the first wall - 1 MW/m^2

Table 4. Nuclide composition of uranium accumulated in thorium HTR blanket, %

Isotope	Irradiation time, days				
	200	400	600	800	1000
^{232}U	1.5	2.86	4.22	5.50	6.70
^{233}U	98.2	96.9	95.3	93.9	92.5
^{234}U	$1.5\cdot10^{-3}$	$3.1\cdot10^{-3}$	$4.8\cdot10^{-3}$	$6.3\cdot10^{-3}$	$7.7\cdot10^{-3}$
^{235}U	$0.2\cdot10^{-5}$	$0.9\cdot10^{-5}$	$2.0\cdot10^{-5}$	$3.6\cdot10^{-5}$	$5.6\cdot10^{-5}$

1.4. ^{231}Pa-^{232}U-^{233}U Chain of Nuclide Transformations with Gradual Perfection of Neutron-Multiplying Properties

In traditional (U-Pu) and (Th-U) fuel cycles, nuclides with good neutron-multiplying properties (^{233}U, ^{235}U, ^{239}Pu, ^{241}Pu) and with neutron-absorbing properties (^{234}U, ^{236}U, ^{240}Pu, ^{242}Pu) alternate with each other. That is why neutron irradiation of traditional fuels results in depletion of fissile nuclides and accumulation of neutron-absorbing nuclides that reduces the reactor criticality and decreases fuel burn-up. In contrast to this situation, if ^{231}Pa produced in thorium HTR blanket is introduced into nuclear fuel composition, then the following chain of consecutive nuclide transformations will be initiated with gradual perfection of neutron-multiplying properties: ^{231}Pa is a neuron-absorbing poison; ^{232}U is a fissionable nuclide with moderate neutron-multiplying properties and, at last, ^{233}U is a well-fissile nuclide with excellent neutron-multiplying properties. The unique feature of this chain consists in the presence of two coupled well-fissile nuclides: ^{232}U and ^{233}U.

^{231}Pa →	^{232}U →	^{233}U →
Burnable poison	**Moderately fissionable nuclide**	**Well-fissile nuclide**

The role of ^{231}Pa as a burnable poison is illustrated by Fig. 4 and Table 5 which show micro cross-sections of radiative neutron capture for ^{231}Pa, for one well-known burnable poison ^{157}Gd and for main fertile isotopes ^{238}U and ^{232}Th.

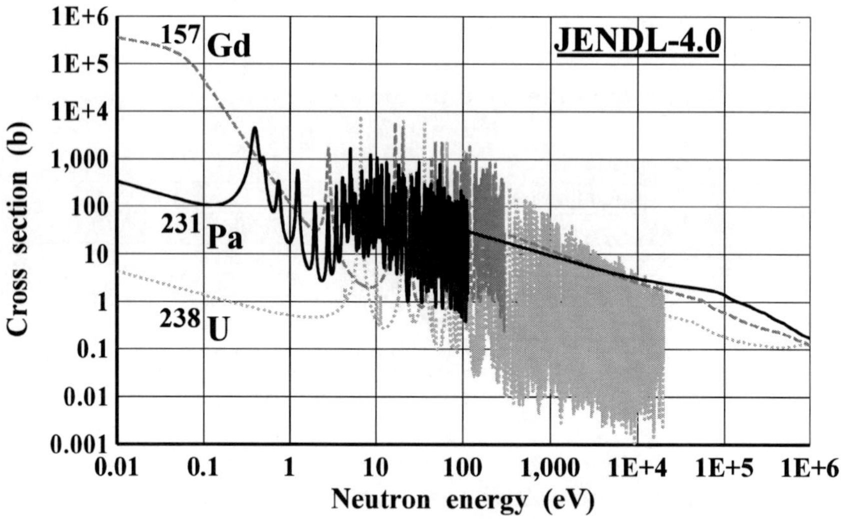

Figure 4. Energy dependence of radiative neutron capture cross-sections for ^{157}Gd, ^{231}Pa and ^{238}U.

Table 5. Radiative neutron capture cross-sections in thermal point and resonance integral

Nuclide	$\sigma_c(V_n$=2200 m/s), barns	Resonance integral of radiative neutron capture, barns
^{157}Gd	253254	784
^{231}Pa	202	540
^{238}U	2.68	275
^{232}Th	7.34	83.8
^{230}Th	23.4	1042
^{232}U	$\sigma_c/\sigma_f = 75.4/76.5$	I_γ/I_f=172/355

As is seen, resonance integrals of radiative neutron capture by ^{231}Pa and ^{157}Gd are comparable with each other while ^{231}Pa is a strong absorber of thermal neutrons: 202 barns in thermal point that is much larger than appropriate micro cross-sections of fertile isotopes ^{238}U and ^{232}Th (2.68 barns and 7.34 barns, respectively), but considerably lower than those of burnable poison ^{157}Gd.

Therefore, even several percentages of ^{231}Pa being introduced into fresh fuel composition can decrease initial reactivity margin thanks to its large radiative neutron capture cross-sections. However, ^{231}Pa is burning-up slower than ^{157}Gd because of relatively smaller radiative neutron capture cross-sections. The slower burning-up of ^{231}Pa may be estimated as a positive effect because stabilization of neutron-multiplying properties can be maintained up to relatively higher values of fuel burn-up.

Thus, analysis of nuclear data allowed us to expect that isotope ^{231}Pa being introduced into nuclear fuel composition will act on time-dependent evolution of neutron-multiplying properties in a way similar with the action induced by well-known burnable poison ^{157}Gd that is widely used in light-water reactors. However, the effects induced by ^{231}Pa as a burnable poison can differ substantially from those induced by ^{157}Gd.

Firstly, neutron absorption cross-sections of ^{231}Pa are not as large as those of ^{157}Gd. Therefore, ^{231}Pa burns slower, and its stabilization effect on neutron-multiplying properties of fuel composition continues acting substantially longer than that of ^{157}Gd. At the same time, neutron absorption cross-sections of ^{231}Pa are substantially higher as compared with those of ^{232}Th. Therefore, relatively small fraction of ^{231}Pa should be introduced into fuel composition to obtain the aforementioned positive effect.

Secondly, it is important to understand that positive effects produced by ^{231}Pa as a burnable poison in fuel composition of thermal reactors differ utterly from those produced by traditional burnable poisons currently used in LWR fuel compositions. Incineration of ^{231}Pa is a self-supporting process in respect to neutron balance of the chain fission reaction and even a neutron-abundant process in high-energy range [11]. The neutrons absorbed by ^{231}Pa will come back to the chain fission reaction through fissions of daughter nuclides. The first of them ^{232}U is a moderately fissionable nuclide within thermal and intermediate energy ranges. The term "moderately fissionable" means here that fission cross-sections of ^{232}U are substantially lower than those of ^{235}U while fission cross-sections and radiative neutron capture cross-sections of ^{232}U are comparable with each other within thermal and resonance energy ranges (Fig. 5).

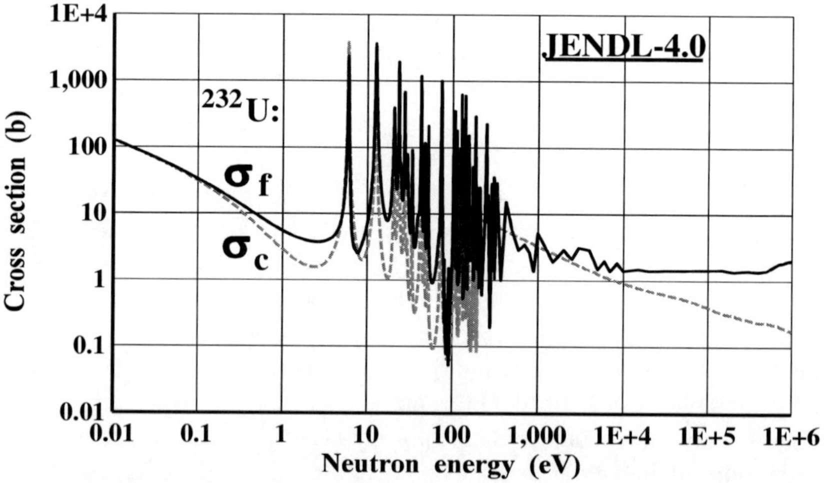

Figure 5. Energy dependence of ^{232}U fission and radiative capture cross-sections.

These nuclear properties of ^{232}U define the lower quantity of excess neutrons per one neutron absorbed by ^{232}U than that of ^{235}U (Fig. 6). However, radiative neutron capture reaction of ^{232}U does not remove this nuclide from neutron balance because it transforms into ^{233}U, widely known and highly efficient fissile nuclide. Fission of ^{233}U can return the previously absorbed neutrons and, in addition, produce large amount of thermal energy. Thus, the aforementioned nuclide transformations lead to the consecutive gradual perfection and long-term stabilization of neutron-multiplying properties during full lifetime of the reactor fuel. Long-lived thorium isotope ^{230}Th (ionium, $T_{1/2} = 7.54 \cdot 10^4$ years) can be produced through the threshold $^{232}Th(n,3n)$

channel. Energy dependencies of radiative neutron capture cross-sections are shown in Fig. 7 for ^{230}Th, ^{232}Th and ^{231}Pa. As is seen, ^{230}Th is a relatively stronger absorber of thermal neutrons than ^{232}Th but ^{230}Th is not as strong neutron absorber as ^{231}Pa. However, ^{230}Th is characterized by the largest value of resonance capture integral.

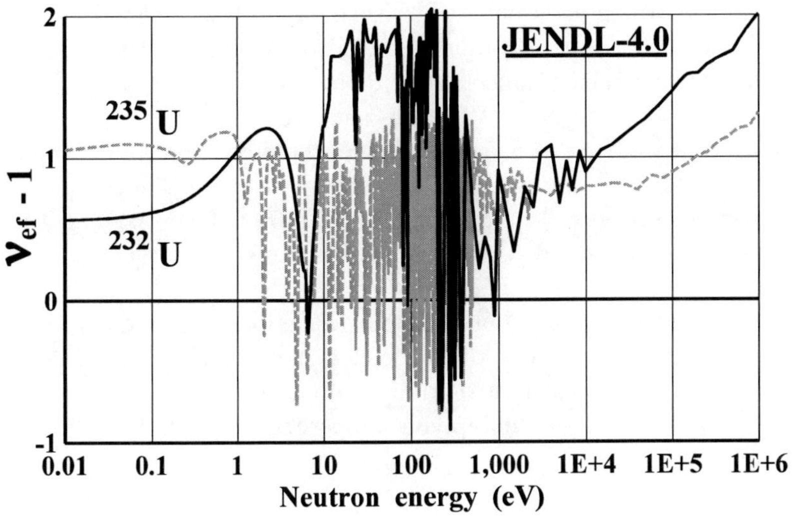

Figure 6. Energy dependence of excess neutron quantity per one neutron absorbed by ^{232}U or ^{235}U.

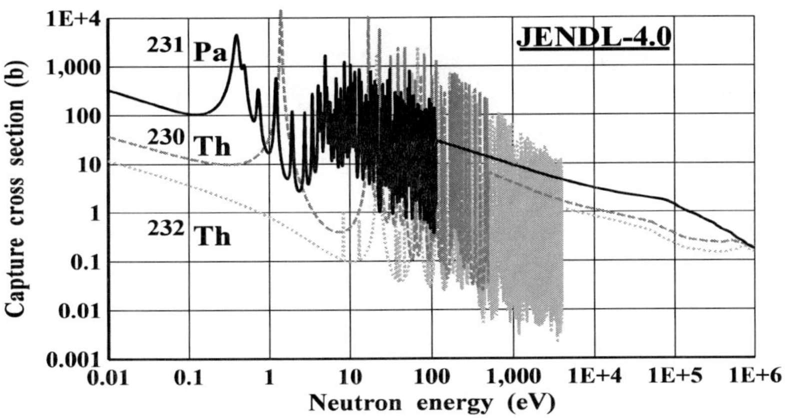

Figure 7. Energy dependence of radiative neutron capture cross-sections for ^{230}Th, ^{231}Pa and ^{232}Th.

Radiochemical reprocessing of thorium irradiated in HTR blanket leaves ^{230}Th together with main thorium isotope ^{232}Th. Thus, ^{230}Th can appear again in thorium HTR blanket for repeat neutron irradiation and transformation into ^{231}Pa by neutron capturing. By the way, ^{230}Th is produced through the threshold ^{232}Th(*n*,3*n*) channel followed by appearance of two additional neutrons, ^{231}Pa is produced through the threshold ^{232}Th(*n*,2*n*) channel followed by appearance of one additional neutron, i.e. both these channels are able to breed neutrons. It seems the most reasonable to place ^{230}Th in peripheral regions of thorium HTR blanket for more effective neutron absorption and so for more intense production of ^{231}Pa, ^{232}U and ^{233}U.

Long-lived isotope ^{230}Th is the third decay product of ^{238}U. Therefore, uranium and mixed uranium-thorium natural ores contain isotope ^{230}Th. So it seems more reasonable to load thorium extracted from the proper deposits into thorium HTR blanket (^{230}Th/U ratio in equilibrium state of ^{238}U decay family is near to 20 ppm).

Thus, the aforementioned nuclide transformations can gradually improve neutron-multiplying properties of the loaded fuel. Such an improvement in competition with accumulation of fission products can form the conditions needed to stabilize neutron-multiplying properties during full lifetime of the reactor fuel.

2. Physical Characteristics of LWR with Ultra-Long Fuel Lifetime

The computer code SCALE-4.3 [12], initially created for the LWR licensing process, was used here for some numerical evaluations. By using the governing module SAS2H from the code SCALE-4.3, time-dependent evolution of neutron-physical parameters in the process of fuel burn-up was calculated for one-dimensional infinite lattice of fuel rods. Neutron transport computations were carried out with application of P_3S_8-approximation and with the number of inner iterations needed to evaluate K_∞ with accuracy of 10^{-5}. Fuel compositions were re-calculated for every 4-year time interval. Calculations of fuel burn-up took into consideration eighty-three fission products selected from their maximal effects on the reactor criticality. The 44-group neutron data library was formed from the evaluated nuclear data file ENDF/B-IV by the computer code AMPX [13]. The number of secondary neutrons produced by ^{232}U fission reactions was corrected according to the recommendations given in Ref. 14.

2.1. Neutron-Physical Validation for Achievability of High Fuel Burn-up

Time-dependent evolutions of neutron multiplication factor in infinite lattice of VVER-type reactor are shown in Fig. 8 for oxide fuel of different compositions. If standard uranium oxide fuel enriched up to 4.4% ^{235}U is used, then neutron multiplication factor decreases monotonously from $K_\infty = 1.38$ to about unity at fuel burn-up of 4.2% HM (curve *1*). Practically the same result was obtained for oxide ^{232}Th-^{233}U fuel initially containing 4% ^{233}U (curve *2*). Time-dependent evolution of neutron multiplication factor changes drastically when ^{231}Pa, the starting nuclide for the coupled pair of fissile nuclides ^{232}U-^{233}U is introduced into fresh fuel composition. Time-dependent evolutions of neutron multiplication factor were analyzed for four nuclide compositions of fresh oxide fuel with different contents of ^{231}Pa, ^{233}U and ^{232}Th.

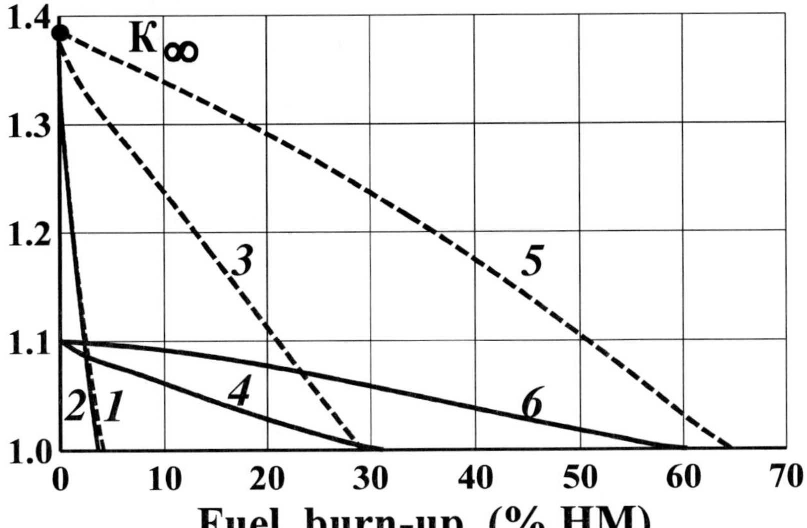

Figure 8. Dependence of neutron-multiplying factor on fuel burn-up for different fuel types:

1 – $(4.4\%^{235}U+95.6\%^{238}U)O_2$;

2 – $(4\%^{233}U+96\%^{232}Th)O_2$;

3 – $(15\%^{231}Pa+22\%^{233}U+63\%^{232}Th)O_2$;

4 – $(30\%^{231}Pa+19\%^{233}U+51\%^{232}Th)O_2$;

5 – $(56\%^{231}Pa+44\%^{233}U)O_2$;

6 – $(71\%^{231}Pa+29\%^{233}U)O_2$.

If 15% ^{231}Pa and 22% ^{233}U (curve *3*) was introduced into fresh fuel composition, then initial value of neutron multiplication factor (K_∞ = 1.38) practically coincides with that for the standard uranium oxide fuel. However, if fractions of ^{231}Pa and ^{233}U in fresh fuel composition were elevated up to 30% and 19%, respectively (curve *4*), then initial value of neutron multiplication factor (K_∞ = 1.10) decreases substantially but this value is high enough to make up for neutron leakage and provide reliable control of the reactor operation. In the process of fuel burn-up, neutron-multiplying properties of both Pa-containing fuel types weakened monotonously down to K_∞ = 1.0 when fuel burn-up reached a value of 30% HM. These compositions of Pa-containing fuels differ advantageously from previous ones by relatively smoother dependency of the reactor criticality on fuel burn-up. These fuel compositions demonstrated stable neutron-multiplying properties in the process of fuel burn-up thanks to the growing contribution from the nuclide transformation chain ^{231}Pa-^{232}U-^{233}U.

The curves 5 and 6 presented in Fig. 8 illustrate seemingly the variants with maximal achievable values of fuel burn-up under the adopted assumptions on the reactor type and fuel types [14]. In these variants of fresh fuel compositions thorium is absent at all while ^{231}Pa fractions were determined basing on the requirement that initial values of neutron multiplication factor would be equal to 1.4 and 1.1, respectively. As ^{231}Pa plays roles of burnable neutron poison and neutron predecessor for the coupled pair of fissile nuclides ^{232}U-^{233}U, it becomes possible to reach the highest values of fuel burn-up (60-65% HM) prior to neutron multiplication factor reduces to unity (K_∞ = 1.0).

The curves shown in Fig. 8 refer to fuel compositions with dominant content of thorium. As we did not find a published information about density of protactinium dioxide, we had to take density of Pa-containing compositions equaled to density of thorium dioxide (9.6 g/cm^3).

The concepts of advanced LWR with supercritical coolant parameters (SCLWR) are often considered as a perspective for further perfection of Russian VVER-type reactors. The supercritical VVER is characterized by the lowered coolant density and, as a consequence, by the harder neutron spectrum, i.e. the spectrum shifted towards higher neutron energies. Under these conditions, introduction of ^{231}Pa into fuel composition as an initial isotope for the nuclide transformation chain ^{231}Pa-^{232}U-^{233}U results in no additional neutron consumption. On the contrary, Pa-containing fuel compositions are characterized by a positive neutron balance [11]. The curves shown in Fig. 9 confirm this effect by demonstrating long-term stabilization of neutron-multiplying properties during the process of fuel burn-up in SC-VVER lattice.

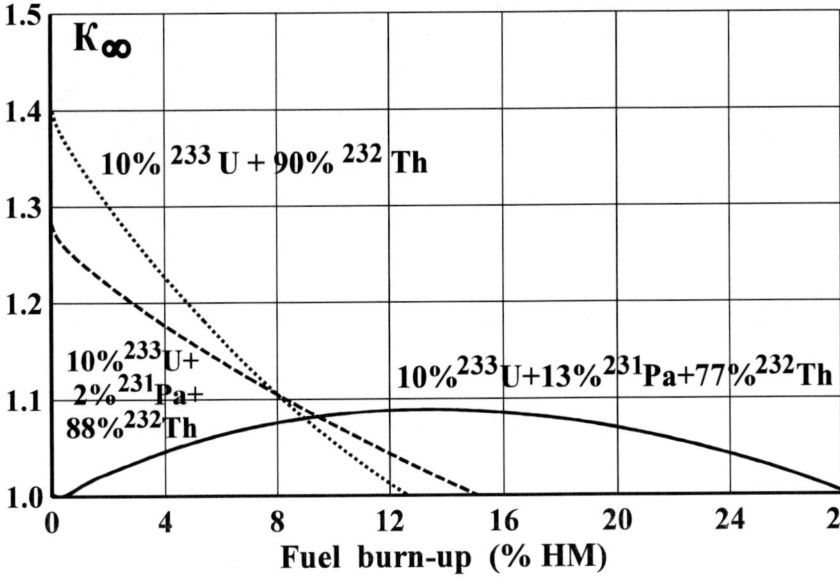

Figure 9. Effect of ^{231}Pa introduction on fuel burn-up (γ_{water} = 0.1 g/cm^3).

In the variant of the standard uranium oxide fuel enriched up to 4.4% ^{235}U, cladding of fuel rods was made of zirconium-based alloy (thickness - 0.65 mm) while in all other variants cladding of fuel rods was made of stainless steel (thickness – 0.4 mm) [16]. As it follows from Figs. 8 and 9, neutron abundance in the chain ^{231}Pa-^{232}U-^{233}U is so large that even the deterioration of neutron balance caused by using steel cladding instead of zirconium cladding in fuel rods allows fuel burn-up to reach the higher values. Besides, the presence of ^{231}Pa shifts thermal and epithermal neutrons towards higher-energy range where neutron absorption by stainless steel decreases (Fig. 10).

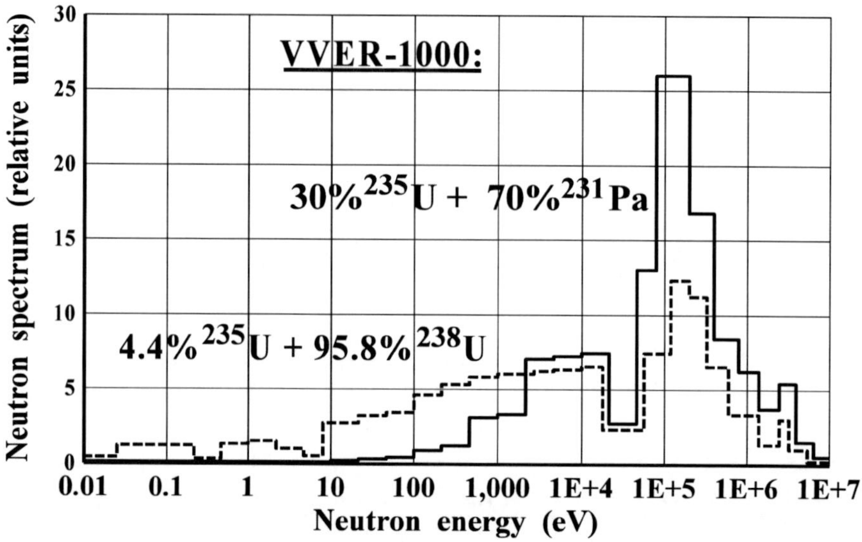

Figure 10. Effect of ^{231}Pa introduction on neutron spectrum.

2.2. Advantages of Higher Fuel Burn-up

The following benefits can be received from the fuel compositions with stabilized neutron-multiplying properties up to high values of fuel burn-up. The first of all, scope of the technological operations related with fuel fabrication, transportation and refueling can be radically reduced. For example, transfer from the traditional LWR fuel with fuel burn-up at the level of 4-6% HM to non-traditional fuel compositions with acceptable values of fuel burn-up at the level 30% HM can reduce the scope of the aforementioned operations by a factor of 5-7. In addition to some economic benefits, the reduced number of fuel transportations leads to radical decreasing the operations which could result in fissile nuclides switching over to non-energy purposes. This advantage can enhance NFC protection against unauthorized proliferation of weapon-grade nuclear materials.

One else factor that can enhance proliferation protection of uranium-based fuel consists in accumulation of ^{232}U, the isotope which can complicate substantially uranium switching over. Isotope ^{232}U is an intense and long-acting source of α-particles ($T_{1/2}$ = 68.9 years). Intensity of high-energy γ-radiation emitted by ^{232}U decay products increases with time because of their gradual accumulation. Also, intense α-radiation can complicate applying the existing gas centrifuge technology of isotope separation for removal of ^{232}U. Besides, α-decays of ^{232}U lead to such intense heat generation (830 W/kg) that, if such a fuel will be switched over, is able to overheat key components of nuclear explosive device (NED). High-energy γ-radiation emitted by ^{232}U decay products in the NED charge with remarkable ^{232}U content will require a special technology for remote handling with the NED charge. Isotope ^{232}U is an intense neutron source from spontaneous fission reactions (1300 n/s·kg) [17]. Moreover, α-particles emitted by ^{232}U are able to generate neutrons in (α,n)-reactions with nuclei of light elements which are always present in fissile materials as small impurities. As is known [17], energy yield of NED with such a nuclear charge will be substantially lowered because of premature initiation of the chain fission reaction (so called, pre-detonation regime).

2.3. Spent Fuel Compositions

As is seen from Table 6, spent fuel having the following fresh composition - 30%^{231}Pa+19%^{233}U+51%^{232}Th – and being characterized by favorable time-dependent evolution of the reactor criticality (curve 4 in Fig. 8) contains about 61% ^{232}U in uranium fraction at the end of the fuel lifetime.

This means that specific heat generation rate of uranium containing well-fissile isotope ^{233}U, very attractive material for nuclear proliferation, is equal to 503 W/kg, i.e. approximately by 48 times higher than that in reactor-grade plutonium (10.5 W/kg) and by 219 times higher than that in weapon-grade plutonium (2.3 W/kg) [18]. The intense heat generation can substantially complicate any non-energy use of such uranium. Surely, at the beginning of the fuel lifetime, the heat generation rate is relatively low in value (fresh fuel generates insignificant amount of heat). In order to enhance proliferation resistance of fresh fuel, a certain quantity of ^{232}U may be introduced into fresh fuel composition. Multiple recycle of regenerated fuel provides the enhancement in a natural manner.

Table 6. Neutron-physical parameters of VVER-type reactor

Fuel type		Dioxide (^{231}Pa+^{233}U+^{232}Th)			Uranium dioxide
Fresh fuel composition, %	^{231}Pa	0	30	71	4.4% ^{235}U 95.6% ^{238}U
	^{233}U	4	19	29	
	^{232}Th	96	51	0	
Fuel burn-up, % HM		3.6	31	60	4.2
Spent fuel composition, %	^{232}Th	93.45	45.07	0.08	1.29% ^{235}U
	^{231}Pa	0.0075	6.28	11.47	0.53% ^{236}U
	^{232}U	0.0045	11.30	22.16	92.87% ^{238}U
	^{233}U	2.47	5.01	3.50	0.66% ^{239}Pu
	^{234}U	0.39	1.74	1.95	0.23% ^{240}Pu
	^{235}U	0.068	0.49	0.65	0.16% ^{241}Pu
	^{236}U	0.006	0.10	0.19	0.05% ^{242}Pu
Beginning of fuel lifetime (0)					
$dK_\infty(0)/dt_{fuel} \cdot 10^{-5}$, 1/°C		-3.3	-2.9	-2.4	-2.9
$dK_\infty(0)/dt_{coolant} \cdot 10^{-4}$, 1/°C		-2.7	+0.61	+1.8	-7.2
$\Delta K_\infty^{work-300^\circ C}(0)$		-0.0300	-0.0260	-0.0176	-0.0270
$\Delta K_\infty^{300^\circ C-50^\circ C}(0)$		-0.0187	-0.0159	+0.0059	-0.0601
End of fuel lifetime (T)					
$dK_\infty(T)/dt_{fuel} \cdot 10^{-5}$, 1/°C		-2.5	-2.0	-0.74	-2.7
$dK_\infty(T)/dt_{coolant} \cdot 10^{-4}$, 1/°C		-2.1	+2.4	+6.9	-7.0
$\Delta K_\infty^{work-300^\circ C}(T)$		-0.0230	-0.0165	-0.0032	-0.0251
$\Delta K_\infty^{300^\circ C-50^\circ C}(T)$		-0.0146	+0.0084	+0.0487	-0.0524

2.4. Reactivity Effects

Dependencies of neutron multiplication factor on temperature and density of fuel and coolant were calculated for infinite lattice of fuel rods in LWR of VVER-type fueled with uranium and (^{231}Pa-^{232}Th-^{233}U) dioxide fuel (Table 6). As is seen, if fuel consists mainly of fertile nuclides (^{238}U or ^{232}Th), then all the reactivity coefficients are negative, i.e. comfortable for the reactor safety. If 30% ^{231}Pa is introduced into fresh fuel composition, then coolant temperature reactivity coefficient $dK_\infty/dt_{coolant}$ (that consists of two components due to changes in both temperature and density of coolant) becomes positive, i.e. uncomfortable for the reactor safety both at the beginning and at the end of fuel lifetime. At the end of fuel lifetime, when the reactor starts up after refueling, temperatures of fuel and coolant are elevated from the cold state (50 °C) up to the hot state (300°C). Appropriate reactivity effect is positive too, although low in value. If ^{231}Pa content in fresh fuel composition increases up to 71%, then Doppler reactivity coefficient dK_∞/dt_{fuel} remains negative but decreases in its absolute value. Besides, coolant temperature reactivity coefficient and the reactivity effect caused by elevation of fuel and coolant temperatures from the cold state to the hot state became positive, both for fresh and spent fuel compositions.

In general, introduction of ^{231}Pa into fresh fuel composition can lead to the decrease (in absolute values) of negative (favorable) safety parameters and even to the appearance of positive (unfavorable) safety parameters in VVER-type lattices of fuel rods. However, we considered here infinite lattice of fuel rods, i.e. neutron leakage from the reactor core was not taken into account. Proper accounting for finite dimensions of the reactor core can shift all the safety parameters towards their negative (favorable) values. It is appropriate to remind here that transportable and transport reactors are characterized, as a rule, by relatively smaller sizes in comparison with large energy reactors of VVER-type.

2.5. Feasibility of High Fuel Burn-Up

One technical question naturally arises: is it feasible to reach high and ultra-high values of fuel burn-up? As it was already mentioned above, some advanced fuel rods containing vibration-compacted oxide fuel have reached fuel burn-up about 32% HM [16].

Non-traditional oxide fuel composition under consideration here contains significant amount of ^{231}Pa. Under neutron irradiation in the reactor core ^{231}Pa converts partially into ^{232}U, initial isotope of natural decay chain ending by stable lead isotope ^{208}Pb. This chain includes six α-decays resulting in helium accumulation in fuel. If addition of gaseous helium to gaseous fission products reaches a remarkable value, then the problem of keeping integrity of fuel cladding becomes more complicated to solve because of higher internal pressure. Numerical evaluations were carried out under assumption that gaseous fission products contain mainly inert gases (xenon, krypton), tritium (product of ternary fissions) and helium as an addition from α-decays. The calculations performed with application of the computer code SCALE-4.3 [12] gave the following results on helium accumulation in fuel pellets for different values of fuel burn-up. If fresh fuel contains 30% ^{231}Pa, then helium addition to total amount of gaseous fission products is equal to 96% at fuel burn-up of 31% HM. If fresh fuel contains 71% ^{231}Pa, then helium addition increases up to 242% at fuel burn-up of 60% HM. It is evident that helium accumulation in fuel pellets at high and ultra-high values of fuel burn-up can complicate significantly the problem of keeping fuel cladding intact. Therefore, it seems reasonable to apply periodically any DUPIC-like technology [19] in order to remove gaseous substances from spent fuel composition and, then, continue the reactor operation.

Table 7. Some parameters of fuel, stainless steel and neutron field by the end of fuel lifetime

Fuel type		(^{231}Pa-^{233}U-^{232}Th) dioxide			Uranium dioxide
Fresh fuel composition, %	^{231}Pa	0	30	71	4.4% ^{235}U 95.6% ^{238}U
	^{233}U	4	19	29	
	^{232}Th	96	51	0	
Fuel burn-up, % HM		3.6	31	60	4.2
Displacements per atom		6.5	66	128	7.9
Neutron fluence, 10^{23} n/cm^2, $E_n > 100$ keV		0.12	1.14	3.01	0.13

Intense neutron irradiation of structural materials in the reactor operation process can induce deterioration of their strength properties and, as ultimate result, their destruction. There is a viewpoint now that some marks of stainless steels currently used as structural materials of fuel rods and fuel assemblies are able to maintain their strength properties without destruction up to the damage dose of 180 displacements per atom [20]. If this assumption is correct, then the fuel composition containing 71% ^{231}Pa may be used up to fuel burn-up about 60% HM without any renewal of fuel cladding (Table 7). According to information taken from Ref. 21, fuel rods with stainless steel cladding are able to withstand the neutron fluence up to $4 \cdot 10^{23}$ n/cm^{2}$\cdot$($E_n > 100$ keV). Our evaluations demonstrated that, for all variants of fresh fuel composition under analysis here, the values of high-energy neutron fluence are well below 180 displacements per atom even in the case of the highest fuel burn-up (60% HM). However, if necessary, a thermo-mechanical DUPIC-like technology may be used for re-fabrication of fuel rods and continuation of the reactor operation.

2.6. Extension of Fuel Lifetime up to 50 Years

Numerical studies allowed us to evaluate some physical parameters (Table 8) of those LWR which may be referred to as the "Black-Box" reactors, i.e. the reactors in which fuel lifetime is comparable or even equal to the reactor lifetime (about 50 years). Table 8 presents also, for comparison, appropriate characteristics of VVER-type reactor with standard composition of uranium dioxide fuel (4.4% ^{235}U).

Table 8. Some parameters of VVER-type and "Black-Box"-type reactors

Parameters	VVER, UO_2	"Black-Box", oxide fuel			
Fresh fuel composition	4.4% ^{235}U	30% ^{231}Pa, 19% ^{233}U, 51% ^{232}Th		71% ^{231}Pa, 29% ^{233}U	
Fuel density, g/cm^3	10.3	9.6	9.6	9.6	9.6
Fuel burn-up, % HM	4.2	31	31	60	60
Heat generation, MW/m^3	113	78	39	113	75
Fuel lifetime, years*	2.33	25	50	33	50
Displacements per atom (stainless steel)	–	66	66	128	128

* - without accounting for duration of the refueling process.

As is seen, 50-year fuel lifetime can be reached only at the values of fuel burn-up above 31% HM but at the expense of a certain decreasing heat generation rate in the reactor core. If target value of fuel burn-up is equal to 31% HM, then heat generation rate should be reduced by a factor of 2.9, and by a factor of 1.5 at 60% HM of target fuel burn-up. By the way, reduced values of heat generation rate are widely used now by developers of advanced nuclear reactors with upgraded operation safety [21]. The calculations were carried out for reduced values of heat generation rate, and the results demonstrated that time-dependent evolution of the reactor criticality lengthened only in a time scale as compared with the calculations for standard value of heat generation rate. Although the damage dose of steel cladding is rather high (128 displacements per atom at fuel burn-up of 60% HM) but this value is regarded now as a high but quite achievable one in the nearest future.

As it was already mentioned above, long lifetimes of ^{231}Pa-containing fuel lead to large amount of gaseous substances thanks to intense helium production by α-decays of ^{232}U and its daughter products. Numerical evaluations revealed that, after 50-year fuel lifetime, amounts of gaseous substances increased on 266% and 340% for fresh fuel compositions containing 30% ^{231}Pa and 71% ^{231}Pa, respectively. This feature can be a serious problem that should be taken into consideration later on. Probably, the problem could be resolved by using DUPIC technology for periodical removal of gaseous substances from spent fuel composition followed by the repeat use of reprocessed and re-fabricated fuel in the reactor core. Release of gaseous substances from fuel rods into coolant of primary circuit is not excluded too.

3. Concept of International Closed NFC

Within the frames of the GNEP (Global Nuclear Energy Partnership) initiative [22, 23] the US President has proposed the following international form of the closed NFC: nuclear power plants (NPP) all over the world must be fed by natural or regenerated uranium only; excess plutonium must be completely utilized in the dedicated reactors – plutonium burners; the plutonium burners must be placed on the territory of the dedicated international nuclear technology centers. Hopefully, such a globalization of the closed NFC is able to provide non-proliferation of weapon-grade nuclear materials. However, there are the following three negative aspects in the GNEP concept.

Firstly, the GNEP concept is oriented on the wide use of relatively cheap uranium resources but foresees a possibility of principle to use secondary, plutonium-bearing fuel. Thus, either the international centers must contain all stages and all enterprises of plutonium NFC or the international centers must provide so reliable proliferation protection of plutonium-bearing fuel that the fuel may be safely and securely utilized at NPP outside of the centers. It is evident that both these alternatives are very difficult to perform.

Secondly, as is known, one ton of uranium being burnt-up in power LWR can produce about one third ton of plutonium. Complete energy utilization of plutonium in the international centers means that about 30% of global nuclear power must be concentrated in the centers. It would be an unprecedented concentration of NPP in one place even for the existing scale of global nuclear power [24].

Thirdly, the wide, already global-scale application of well-developed gas-centrifuge technology for isotope enrichment of natural uranium has led to the real fact that low enriched uranium (LEU) used as a main fuel type for NPP all over the world became the most attractive nuclear material for potential proliferators who can easily re-enrich LEU up to the weapon-grade level [25,

26]. Thus, fresh LEU-based fuel for NPP must be also proliferation protected with a high degree of reliability.

These negative peculiarities of the GNEP concept may be avoided by introducing HTR with thorium blanket into structure of the international nuclear technology centers. We propose the following scheme for a global-wide closed NFC that includes the international nuclear technology centers responsible for spent fuel reprocessing and park of LWR responsible for energy production throughout the world (Fig. 11).

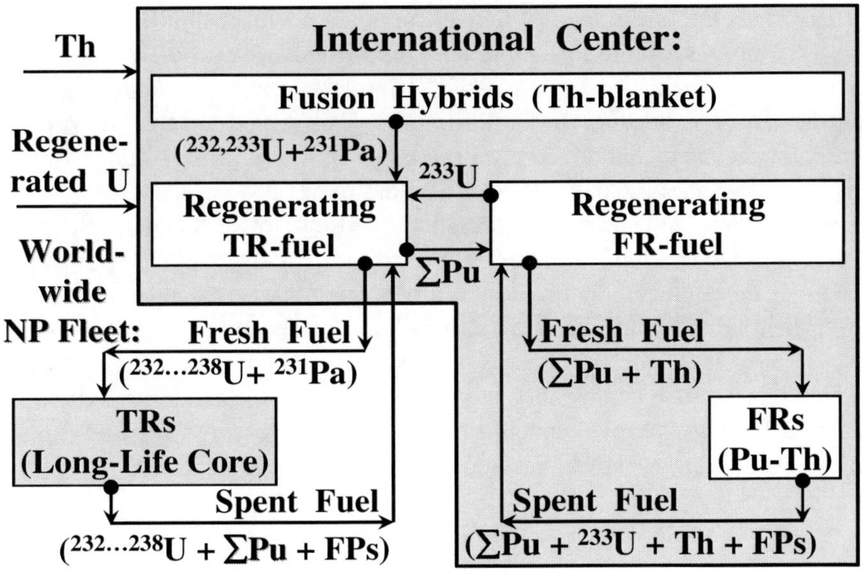

Figure 11. Closed NFC with joint utilization of regenerated uranium and thorium.

The closed NFC presented in Fig. 11 is based on the following assumptions. The international nuclear technology centers are fed by regenerated uranium and thorium. Fast reactors in the centers are fed by mixed (Pu-Th) fuel for plutonium conversion into ^{233}U. HTR with thorium blankets are used to produce (^{231}Pa-^{232}U-^{233}U) mixture. Thermal reactors, outside of the centers, are fed by multi-isotope uranium-protactinium $^{232\text{-}236,238}$U-^{231}Pa fuel which may be fabricated by mixing, in appropriate proportions, of the following components: regenerated uranium ^{234,235,236,238}U extracted from spent LWR fuel assemblies, uranium-protactinium mixture ^{231}Pa-^{232}U-^{233}U extracted thorium HTR blankets and ^{233}U extracted from spent (Pu-Th) fuel of fast reactors. So, no any isotope separation technology is used in the scheme.

The presence of significant ^{231}Pa quantity in the mixed fuel can ensure stable enough neutron-multiplying properties up to high fuel burn-up. The presence of ^{231}Pa improves neutron balance and shifts neutron spectrum towards higher energy range. These effects allow the LWR developers to replace zirconium claddings of fuel rods by steel ones. Thus, it becomes feasible to reach high and ultra-high values of fuel burn-up in LWR.

The mixed fuel is characterized also by enhanced protection against unauthorized proliferation. In the fuel composition well-fissile isotopes ^{233}U and ^{235}U are surrounded from both sides by lighter and heavier uranium isotopes – neutron absorbers. Moreover, uranium contains significant fraction of ^{232}U. Plutonium build-up is strongly suppressed because of reduced ^{238}U content in the fuel. The presence of ^{236}U in the uranium fraction initiates neutron reactions leading to build-up of ^{238}Pu, which is able to protect plutonium against unauthorized proliferation [17]. In this sense, ^{238}Pu plays the same role for plutonium that ^{232}U plays for uranium.

In the proposed scheme of international closed NFC full energy potential of uranium and thorium is completely used, i.e. fuel base of nuclear power system becomes practically unlimited. It is noteworthy here that share of HTR power in total nuclear power system will be rather small because energy released in one fusion reaction is substantially lower than that in fission reaction of nuclear reactors and spallation reaction of accelerator-driven cores [27]. That is why the involvement of HTR into the closed NFC could impart the following principal advantages to nuclear power system [28-30]:

- After a relatively short time, the global nuclear power system can refuse natural uranium mining (as well as exploring and mastering of new uranium deposits) and use the regenerated uranium only. It is a well-known fact that amount of uranium in already mined uranium ores, spent uranium-based fuel and depleted uranium (by-product of uranium enriching) is large enough to meet the resource demands of the global nuclear power system during the next several centuries.
- After a relatively short time, the global nuclear power system can refuse a necessity to apply any proliferation-dangerous isotope separation technology for production of nuclear fuel.
- HTR-produced fissile materials are the proliferation-proof materials.
- Low-enriched uranium to be used as a fuel of nuclear power reactors becomes protected against any attempts of its re-enrichment up to the weapon-grade level.

- Uranium remains for a long time period to be a main fuel component of nuclear power reactors, and the component containing both fissile ^{233}U and fertile ^{238}U nuclides.
- Long-life cores of nuclear power reactors can reduce radically the circulation flow of nuclear fuel materials to be reprocessed, enhance potential ability of the reactors to work in autonomous operation mode, shorten drastically transportation flows of highly radioactive materials and upgrade export potential of nuclear power technologies.

CONCLUSION

The following conclusions may be derived from the results presented above:

1) Thorium HTR blanket provides the larger yield of nuclides defining stable neutron-multiplying properties and proliferation protection of advanced fuel compositions per one fusion neutron than uranium HTR blanket does.
2) LWR cell loaded with fuel extracted from thorium HTR blanket can reach high (30% HM) and ultra-high (60% HM) values of fuel burn-up.
3) Reactivity effects of LWR cell loaded with fuel extracted from thorium HTR blanket are only slightly inferior to those of LWR cell loaded with traditional uranium oxide fuel.
4) Analysis of technical difficulties related with reaching high and ultra-high fuel burn-up (fuel swelling, accumulation of gaseous fission products, neutron-induced damages of structural materials) allowed us to hope on a feasibility to solve all these problems.
5) The presence of ^{232}U in uranium fraction can play a significant role in proliferation protection of advanced fuel compositions.
6) The proposed scheme of international closed NFC including HTR with thorium blanket opens a possibility of principle for full energy utilization of already available uranium and thorium. Thus, fuel base of the world-scale nuclear power system becomes practically unlimited.
7) The share of HTR power in total nuclear power system will be rather small because energy released in one fusion reaction is substantially lower than that in fission or spallation reaction.

8) The proposed scheme of international closed NFC is characterized by flexible and easy to re-build structure with the sparing requirements to efficiency of fuel utilization in nuclear power reactors.

ACKNOWLEDGMENT

The chapter was prepared with financial support given by the International Science and Technology Center (http://istc.ru) within the framework of the Responsible Science Program of Sub-Program SB159 “Culture of Nuclear Non-Proliferation” as well as by the Russian Scientific Fund (http://grant.rscf.ru) in frames of project №14-19-00276 “Radical increase of fuel burn-up of nuclear reactors and proliferation protection of fissile materials by introducing hybrid thermonuclear reactors FUSION-FISSION into closed nuclear fuel cycle for energy supply of remote regions”.

REFERENCES

[1] E.G. Kulikov, G.G. Kulikov, E.F. Kryuchkov, A.N. Shmelev. “Achievement of higher LWR fuel burn-up by introducing ^{231}Pa into fuel composition”. – *Nuclear Physics and Engineering,* Volume 4, No. 4, pp. 291–299 (2013).

[2] A.N. Shmelev. “Potential role of controlled thermonuclear fusion in resolving the problems of radioactive wastes and proliferation-protected fuel cycles with increased fuel burn-up”. – *Communications of Higher Schools. Nuclear Power Engineering,* Volume 14, No. 4, pp. 53-59 (1997).

[3] A.N. Shmelev, G.G. Kulikov, E.G. Kulikov, E.F. Kryuchkov, V.A. Apse. “Controlled thermonuclear fusion: potential role in the united (Th-U-Pu) nuclear fuel cycle”. - *Nuclear Physics and Engineering,* Volume 2, No. 2, pp. 101–111 (2011).

[4] H.E.J. Shief, et al. “Measurements of the reaction rate distributions produced in a large thorium cylinder by a central source of DT neutrons. - AWRE 20/77, Atomic Energy Authority, United Kingdom (1977).

[5] Krumbein, M. Lemanska, M.Segev, J.J. Wagschal, A. Yaari. “Reaction rate calculations in uranium and thorium blankets surrounding a central

deuterium-tritium neutron source". - *Nuclear Technology*, Volume 48, pp. 110–116 (1980).
[6] J.W. Weale, H. Goodfellow, M.H. McTaggart, M.L. Mullender. "Measurements of reaction rate distribution produced by a source of 14-MeV neutrons at the center of a uranium metal pile". - *Reactor Science and Technology (Journal of Nuclear Energy, Parts A and B)*, Volume 14, pp. 91–99 (1961).
[7] R.C. Haight, J.D. Lee, J.A. Maniscalco. "Reaction rates in a uranium pile surrounding a 14 MeV neutron source: Calculations of the Weale experiment". - *Nuclear Science and Engineering*, Volume 61, pp. 53–59 (1976).
[8] K.R. Schultz. "A review of hybrid reactor fuel cycle considerations". – Proceedings of the Second Soviet-American Seminar "Fusion-Fission" (March 14 – April 1, 1977, Moscow, Russia), pp. 77–93 (1978).
[9] S.B. Marin, G.E. Shatalov. "Isotope composition of fuel in thorium blanket of hybrid thermonuclear reactor". – *Atomic Energy*, Volume 56, No. 5, pp. 289–291 (1984).
[10] E.O. Adamov, A.V. Dgalavian, A.V. Lopatkin, et al. Conceptual propositions of the strategy for nuclear power development in Russia up to 2100. – Atomic Energy, Volume 112, No. 6, pp. 319-331 (2012).
[11] G.. Kulikov., A. Shmelev, E. Kulikov, V. Apse V. Neutron-physical peculiarities of nuclide chains for actinide isotopes, and proliferation protection of fissionable isotopes. – Proceedings of the Twelfth International Conference on Emerging Nuclear Energy Systems ICENES-2005 (Brussels, Belgium), on CD-ROM, paper 076 (2005).
[12] "SCALE: a modular code system for performing standardized computer analyses of licensing evaluation". - NUREG/CR-0200, ORNL/NUREG/CSD-2/RS, Oak Ridge, USA, (1997).
[13] J. Greene, J.L. Lucius, L.M. Petrie, et al. "AMPX: a modular code system for generating coupled multi-group neutron-gamma libraries from ENDF/B". - ORNL/TM-3706 Oak Ridge, USA, (1976).
[14] A.N. Shmelev, G.G. Kulikov. "About neutron-physical features of modified (denatured) fuel cycles". – *Communications of Higher Schools. Nuclear Power Engineering*, Volume 6, pp. 42–48 (1997).
[15] A.N. Shmelev, M. Saito, V. Artisyuk. "Multi-component self-consistent nuclear energy system: One proliferation resistant aspect". – Proceedings of the Second Annual JNC International Forum "*Peaceful*

Use of Nuclear Energy" (Tokyo, Japan, February 21-22, 2000), pp. 87–95 (2000).

[16] V.B. Ivanov, A.A. Mayorshin, O.V. Skiba, et al. "The utilization of plutonium in nuclear reactors on the basis of technologies developed in SSC RIAR". – Proceedings of the International Conference on Future Nuclear Systems GLOBAL-97 (Yokohama, Japan, October 5-10,. 1997), Volume 2, pp. 1093–1098 (1997).

[17] De Volpi. "Denaturing fissile materials". - *Progress in Nuclear Energy*, Volume 10, No. 2, pp. 161–220 (1982).

[18] J.C. Mark. "Explosive properties of reactor-grade plutonium". - *Science & Global Security*, Volume 4, pp. 111–128 (1993).

[19] M.S. Yang, B.O. Kim, K.W. Song, et al. "Characteristics of DUPIC fuel fabrication technology". - Proceedings of the International Conference on Future Nuclear Systems GLOBAL-97 (Yokohama, Japan, October 5-10,. 1997), Volume 1, pp. 535-537 (1997).

[20] S. Pillon, J. Tommasi, T. D'Alletto, et al. "Current status of the CAPRA programme". – Proceedings of the International Conference ENC-98 (Nice, France, October 25-28, 1998), Volume 3, pp. 645–652 (1998).

[21] M.D. Carelli, D.V. Paramonov, C.V. Lombardy, et al. "IRIS, international new generation reactor". – Proceedings of the Eighth International Conference on Nuclear Engineering ICONE-8 (Baltimore, USA, April 2-6, 2000), ICONE-8447 (2000).

[22] R.A. Krakowski, J.W. Davidson, C.G. Bathke, E.D. Arthur, R.L.JR. Wagner. "Nuclear energy and materials in the 21st century". – The IAEA International Symposium "*Nuclear Fuel Cycle and Reactor Strategies: Adjusting to New Realities"* (Vienna, June 1997), IAEA-SM-346/101 (1997).

[23] P.T. Cunningham, E.D. Arthur, R.L. Wagner, E.M. Jr. Hanson. "Strategies and technologies for nuclear materials stewardship". – Proceedings of the International Conference on Future Nuclear Systems GLOBAL-97 (Yokohama, Japan, October 5-10,. 1997), Volume 1, pp. 720-725 (1997).

[24] A.N. Shmelev, G.V. Tikhomirov, G..G.. Kulikov, et al. "About concept of the international nuclear technology centers for plutonium utilization". - *Communications of Higher Schools. Nuclear Power Engineering*, Volume 4, pp. 81-92 (1998).

[25] A.Yu. Smirnov, V.A. Apse, V.D. Borisevich, et al. "Introduction of regenerated uranium into LWR fuel composition as a protective measure

against nuclear proliferation". - *Communications of Higher Schools. Nuclear Power Engineering*, Volume 4, pp. 3–13 (2011).

[26] P.N. Alexeyev, E.A. Ivanov, V.A. Nevinitsa, et al. "Enhanced proliferation protection for export deliveries of LWR fuel by using regenerated uranium". - *Communications of Higher Schools. Nuclear Power Engineering*, Volume 3, No. 2, pp. 3–9 (2007).

[27] I.S. Grigoriev, E.Z. Meilikhov (editors). *Physical values. Handbook.* – Energoatomizdat, Moscow, USSR, p. 1232 (1991).

[28] A.N. Shmelev, V. Artisyuk, M. Suzuki, M. Saito, Y. Fujii-e. "A possible contribution of various neutron sources (fission, spallation, fusion) in excess neutron generation for SCNES". – Proceedings of the Eighth International Conference ICENES-96 (Obninsk, Russia, June 24-28, 1996), pp. 110–120 (1996).

[29] M. Saito, A. Shmelev, V. Artisyuk, M. Suzuki, Yu. Korovin, Y. Fujii-e. "Contribution of external neutron sources in excess neutron generation in SCNES". – Proceedings of the Second International Symposium GENES-2 (Tsuruga, Japan, October 29-November 1, 1996), pp. 697–706 (1996).

[30] V. Artisyuk, A. Shmelev, M. Suzuki, m. Saito, Y. Fujii-e. "Potential of neutron excess generation by DD-fusion in SCNES". – Proceedings of the Second International Symposium GENES-2 (Tsuruga, Japan, October 29 – November 1, 1996), pp. 721-728 (1996).

INDEX

C

D

E

I

J

K

L

M

N

O

P

Q

R

S

T